EXTERIOR SOUND LEVEL MEASUREMENTS OF OVER-SNOW VEHICLES AT YELLOWSTONE NATIONAL PARK

September 2008
Final Report

Prepared for:
National Park Service
PO Box 168
Yellowstone National Park, WY 82190

Prepared by:
U.S. Department of Transportation
Research and Innovative Technology Administration
John A. Volpe National Transportation Systems Center
Environmental Measurement and Modeling Division, RTV-4F
Cambridge, MA 02142

Notice

This document is disseminated under the sponsorship of the Department of Transportation in the interest of information exchange. The United States Government assumes no liability for its contents or use thereof.

Notice

The United States Government does not endorse products or manufacturers. Trade or manufacturers' names appear herein solely because they are considered essential to the objective of this report.

			Form Approved
# REPORT DOCUMENTATION PAGE			OMB No.

Public reporting burden for this collection of information is estimated to average 1 hour per response, including the time for reviewing instructions, searching existing data sources, gathering and maintaining the data needed, and completing and reviewing the collection of information. Send comments regarding this burden estimate or any other aspect of this collection of information, including suggestions for reducing this burden, to Washington Headquarters Services, Directorate for Information Operations and Reports, 1215 Jefferson Davis Highway, Suite 1204, Arlington, VA 22202-4302, and to the Office of Management and Budget, Paperwork Reduction Project (0704-0188), Washington, DC 20503.

1. AGENCY USE ONLY (Leave blank)	2. REPORT DATE September 2008	3. REPORT TYPE AND DATES COVERED Final
4. TITLE AND SUBTITLE Exterior Sound Level Measurements of Over-Snow Vehicles at Yellowstone National Park		5. FUNDING NUMBERS VX82 - EM355
6. AUTHOR(S) Aaron L. Hastings, Chris J. Scarpone, Gregg G. Fleming, Cynthia S. Y. Lee		
7. PERFORMING ORGANIZATION NAME(S) AND ADDRESS(ES) U.S. Department of Transportation Research and Innovative Technology Administration John A. Volpe National Transportation Systems Center Environmental Measurement and Modeling Division, RTV-4F Acoustics Facility Cambridge, MA 02142-1093		8. PERFORMING ORGANIZATION REPORT NUMBER DOT-VNTSC-NPS-08-03
9. SPONSORING/MONITORING AGENCY NAME(S) AND ADDRESS(ES) Mike Yochim PO Box 168 Yellowstone National Park WY 82190 307-344-2024 Mike_Yochim@nps.gov		10. SPONSORING/MONITORING AGENCY REPORT NUMBER
11. SUPPLEMENTARY NOTES Lead National Park Service Technical Resource: Shan Burson, Grand Teton National Park PO Drawer 170, Moose WY 83012, 307-739-3584, Shan_Burson@nps.gov		
12a. DISTRIBUTION/AVAILABILITY STATEMENT		12b. DISTRIBUTION CODE
13. ABSTRACT (Maximum 200 words) Sounds associated with oversnow vehicles, such as snowmobiles and snowcoaches, are an important management concern at Yellowstone and Grand Teton National Parks. The John A. Volpe National Transportation Systems Center's Environmental Measurement and Modeling Division is supporting the National Park Service with its on-going Winter Use Planning program. As part of this support, acoustic measurements of ten snowcoaches and six snowmobiles were made at the south entrance to Yellowstone National Park from the 26th through the 28th of February 2008. Measurement methodologies were guided by SAE J1161 and SAE J192. There were two primary objectives: 1) to determine which snowcoaches had the Best Available Technology (BAT) with respect to noise emissions, and 2) to determine if there was a significant difference between snowmobile sound levels when tested using two revisions of SAE J192. Based on analysis of the data, it appears that among the snowcoaches, the Yellowstone Expeditions' modified Dodge B530 snowcoach would be an excellent candidate for BAT classification. It was also found that measurements guided by the two revisions of SAE J192 had very similar results.		
14. SUBJECT TERMS Noise Measurement, Parks, Snowmobiles, Snowcoaches, Snow, Sound Propagation, Ground Effects, Integrated Noise Model		15. NUMBER OF PAGES 79
		16. PRICE CODE
17. SECURITY CLASSIFICATION OF REPORT Unclassified	18. SECURITY CLASSIFICATION OF THIS PAGE Unclassified	19. SECURITY CLASSIFICATION OF ABSTRACT Unclassified	20. LIMITATION OF ABSTRACT I

NSN 7540-01-280-5500

Standard Form 298 (Rev. 2-89)
Prescribed by ANSI Std. 239-18
298-102

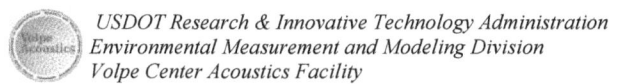
USDOT Research & Innovative Technology Administration
Environmental Measurement and Modeling Division
Volpe Center Acoustics Facility

September 2008

Table of Contents

Section | Page

1. **Introduction** 1
2. **Measurement Setup and Data Collection** 3
   - 2.1. Measurement Site Layout 3
   - 2.2. Measurement Conditions 6
   - 2.3. Vehicle Description 7
   - 2.4. Equipment Description (Acoustics, Meteorology, and Speed) 13
     - 2.4.1. Acoustic System 13
     - 2.4.2. Meteorological Equipment 14
     - 2.4.3. Vehicle Speed Collection (Doppler-Radar Gun) 15
3. **Measurement Methodology** 16
   - 3.1. Methodology in Accordance with SAE International Standard J1161 16
   - 3.2. Methodology in Accordance with SAE International Standard J192 17
   - 3.3. Additional Measurements 17
   - 3.4. Meteorological Measurements 18
4. **Results and Analysis** 20
   - 4.1. Ambient and Meteorological Conditions 20
   - 4.2. Best Available Technology (BAT) for Snowcoach Noise Emissions 20
     - 4.2.1. Sound Level Time History 21
     - 4.2.2. Overall Level (L_{ASmx}) Associated with Each Vehicle 24
     - 4.2.3. Noise vs. Speed Plots 26
     - 4.2.4. Sound Exposure Level (SEL), dB(A) 28
     - 4.2.5. Sound Level Spectra 29
   - 4.3. Comparison of SAE J192 (1985) and SAE J192 (2003) 32
   - 4.4. Snowmobiles at Constant Speed 33
   - 4.5. Ground Attenuation 34
   - 4.6. Snow Bank (Barrier) Effects 35
5. **Summary and Conclusions** 37

Appendix A: Larson Davis Model 824 Sound Level Meter 39
   - A.1 LD824 Settings 39
   - A.2 Check/apply Setup Parameters 39
   - A.3 Setting the Clock 40
   - A.4 Calibration 40
   - A.5 Collect and Monitor Data 40

Appendix B: Sony Model TCD-100 DAT Recorder 41
   - B.1 Sony DAT Settings 41
   - B.2 Check / Apply Setup Parameters 41
   - B.3 Setting the Clock 41
   - B.4 Operation Notes 42

Appendix C: Qualimetrics Transportable Automated Meteorological Station 43
   - C.1 TAMS Unit Setup 43

Appendix D: Measurement Protocol and Logging Procedure 44

Appendix E: Sound Level Time Histories 46

Appendix F: Overall Sound Levels 60
Appendix G: One-Third Octave Band Levels 67
References 77

List of Figures

Section   Page

Figure 1: Sound Level vs. Speed VII
Figure 2. Location of Measurements in the John D. Rockefeller Jr. Memorial Parkway .. 1
Figure 3. Aerial Photo of Measurement Site (44.133, -110.66503). White dots indicate approximate location of microphones.. 4
Figure 4. Measurement Site Sketch – (A) Aerial View, (B) Profile.................................... 5
Figure 5. Measurement of Yellowstone Snow Coach Tours OSV Showing Line-of-Sight Blockage of Tracks due to Accumulated Snow 6
Figure 6. Purpose Built Bombardier B-12 with Skis and Tracks (Alpen Guide) 8
Figure 7. Customized Dodge B350 Van with "Snowbuster" Skis and Tracks (Yellowstone Expeditions).. 9
Figure 8. Customized Econoline 4 x 4 with "Mattracks" Tracks (Rocky Mountain SC Tours).. 9
Figure 9. Customized Van with "Mattracks" Tracks (Yellowstone SC Tours)................. 10
Figure 10. Customized Van with "Mattracks" Tracks (Xanterra 431) 10
Figure 11. Purpose Built Bombardier B-12 with Skis and Tracks (Xanterra 707)........... 11
Figure 12. Purpose built International Bus with "Grip Trax" Tracks (NPS Snowcoach) 11
Figure 13: Arctic Cat TZ1 12
Figure 14. Arctic Cat T660 Snowmobile... 12
Figure 15: Artic Cat Bearcat 13
Figure 16. Acoustic System Setup 14
Figure 17. TAMS Setup 15
Figure 18. SAE J1161 / SAE J192 Measurement Area Layout... 17
Figure 19. Extended Measurement Area Layout 18
Figure 20. Goosewing Diesel Van, Right Side at Low (15 mph) Speed (Feb 26th) 22
Figure 21. Goosewing Diesel Van, Left Side at High (30 mph) Speed (Feb 26th) 22
Figure 22. Goosewing Diesel Van, Left Side at Idle (Feb 26th) 23
Figure 23. Xanterra 707 Idle (Feb 27th) 24
Figure 24. Sound Level vs. Speed 27
Figure 25. Goosewing Diesel Van, Right Side (Feb 26th, 13:52) Spectra for Low Speed 30
Figure 26. Yellowstone Expedition, Right Side (Feb 26, 12:53) Spectra for Low Speed 30
Figure 27. Yellowstone Expedition, Left Side (Feb 26, 14:18) Spectra for High Speed . 31
Figure 28. Xanterra 165, Left Side (Feb 27, 11:50) One-Third Octave Band Spectra for the Microphones 4 and 15 foot above the Snow Cover at 50 feet from the Travel Lane . 36
Figure 29. Alpen Guide, Right Side 15 mph (Feb 26th)... 46
Figure 30. Goosewing Diesel Van, Right Side 15 mph (Feb 26th) 47
Figure 31. Rocky Mt. SC Tours, Right Side 15 mph (Feb 26th)....................................... 47
Figure 32. Yellowstone Expedition, Right Side 16 mph (Feb 26th) 48
Figure 33. Yellowstone Snow Coach, Right Side 16 mph (Feb 26th).............................. 48
Figure 34. Xanterra 431, Right Side 15 mph (Feb 27th) 49
Figure 35. Xanterra 707, Right Side 15 mph (Feb 27th) 49
Figure 36. Xanterra 165, Left Side 15.5 mph (Feb 27th) 50
Figure 37. Goosewing Excursion, Right Side 16 mph (Feb 27th)..................................... 50

Figure 38. NPS Snow Coach, Left Side 15 mph (Feb 28th) 51
Figure 39. Alpen Guide, Right Side 34 mph (Feb 26th) 52
Figure 40. Goosewing Diesel Van, Left Side 28 mph (Feb 26th) 52
Figure 41. Rocky Mt. SC Tours, Left Side 29 mph (Feb 26th) 53
Figure 42. Yellowstone Expedition, Left Side 28.5 mph (Feb 26th) 53
Figure 43. Yellowstone Snow Coach, Right Side 30 mph (Feb 26th) 54
Figure 44. Alpen Guide, Right Side Idle (Feb 26th) 55
Figure 45. Goosewing Diesel Van, Left Side Idle (Feb 26th) 55
Figure 46. Rocky Mt. SC Tours, Left Side Idle (Feb 26th) 56
Figure 47. Yellowstone Expedition, Left Side Idle (Feb 26th) 56
Figure 48. Yellowstone Snow Coach, Left Side Idle (Feb 26th) 57
Figure 49. Xanterra 431, Right Side Idle (Feb 27th) 57
Figure 50. Xanterra 707, Right Side Idle (Feb 27th) 58
Figure 51. Xanterra 165, Left Side Idle (Feb 27th) 58
Figure 52. Goosewing Excursion, Left Side Idle (Feb 27th) 59
Figure 53. NPS Snow Coach, Left Side Idle (Feb 28th) 59
Figure 54. NPS SC, Left Side (Feb 28, 10:18) Spectra for Low Speed 68
Figure 55. Xanterra 707, Right Side (Feb 27, 11:52) Spectra for Low Speed 68
Figure 56. Goosewing Diesel Van, Right Side (Feb 26, 13:52) Spectra for Low Speed . 69
Figure 57. Yellowstone SC Tour, Right Side (Feb 26, 13:21) Spectra for Low Speed 69
Figure 58. Goosewing Excursion, Right Side (Feb 27, 12:32) Spectra for Low Speed ... 70
Figure 59. Xanterra 431, Right Side (Feb 27, 11:27) Spectra for Low Speed 70
Figure 60. Xanterra 165, Left Side (Feb 27, 11:50) Spectra for Low Speed 71
Figure 61. Alpen Guide, Right Side (Feb 26, 12:12) Spectra for Low Speed 71
Figure 62. Rocky Mt, Right Side (Feb 26, 13:43) Spectra for Low Speed 72
Figure 63. Yellowstone Expedition, Right Side (Feb 26, 12:53) Spectra for Low Speed 72
Figure 64. Rocky Mt, Left Side (Feb 26, 15:18) Spectra for High Speed 74
Figure 65. Alpen Guide, Right Side (Feb 26, 14:07) Spectra for High Speed 74
Figure 66. Goosewing Diesel Van, Left Side (Feb 26, 15:16) Spectra for High Speed ... 75
Figure 67. Yellowstone Snow Coach, Right Side (Feb 26, 14:40) Spectra for High Speed
.... 75
Figure 68. Yellowstone Expedition, Left Side (Feb 26, 14:18) Spectra for High Speed . 76

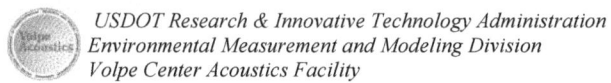

List of Tables

Section   Page

Table 1. Snowcoach Vehicle Description VII
Table 2. Snowmobile Vehicle Description VIII
Table 3. Snowcoach Vehicle Description 7
Table 4. Snowmobile Vehicle Description 7
Table 5. TAMS Measurement Capabilities 15
Table 6. Example Sound Level Calculation for SAE J1161 17
Table 7. Summary of Meteorological Conditions During the Measurements 20
Table 8. L_{ASmx} for Loudest Side of Vehicle at Low Speed, dB(A) 25
Table 9. L_{ASmx} for Loudest Side of Vehicle at High Speed, dB(A) 25
Table 10. L_{Aeq} for Loudest Side of Vehicle at Idle (0 mph), dB(A) 26
Table 11. Average SEL for each Event at Low Speed, dB(A) 28
Table 12. Average SEL for each Event at High Speed, dB(A) 29
Table 13: Differences between the 1985 and 2003 Revisions of SAE J192 32
Table 14: Comparison 33
Table 15. L_{ASmx} for Loudest Side of Vehicle at Low Speed, dB(A) 33
Table 16. L_{ASmx} for Loudest Side of Vehicle at High Speed, dB(A) 33
Table 17. Average SEL for each Event at Low Speed, dB(A) 34
Table 18. Average SEL for each Event at High Speed, dB(A) 34
Table 19. Ground Effect Results for Three Vehicles Measured on February 27th. 35
Table 20. Larson Davis 824 real time analyzer settings. 39
Table 21. Low Speed Measurements used to Generate Final Reported L_{ASmx} Level for each Snowcoach 60
Table 22. High Speed Measurements used to Generate Final Reported L_{ASmx} Level for each Snowcoach 62
Table 23. Idle Measurements used to Generate Final Reported L_{Aeq} Level for each Snowcoach 63
Table 24. Low Speed Measurements used to Generate Final Reported SELs for each Snowcoach 64
Table 25. High Speed Measurements used to Generate Final Reported SELs for each Snowcoach 66
Table 26. Maximum One-Third Octave Band Levels for Select Events at Low Speed (Nominally 15 mph), dB 67
Table 27. Maximum One-Third Octave Band Levels for Select Events at High Speed (Nominally 30 mph), dB 73

USDOT Research & Innovative Technology Administration
Environmental Measurement and Modeling Division
Volpe Center Acoustics Facility

Executive Summary

Sounds associated with Oversnow Vehicles (OSVs), such as snowmobiles and snowcoaches, are an important management concern at Yellowstone and Grand Teton National Parks. The John A. Volpe National Transportation Systems Center's Environmental Measurement and Modeling Division (Volpe Center) is supporting the National Park Service (NPS) with implementation of the Winter Use Planning program (Ref. 1, 2, 3, 4) and supporting National Environmental Policy Act (NEPA) documents, including the 2007 Winter Use Planning / Environmental Impact Statement. As part of this support, the Volpe Center, in cooperation with the NPS, performed acoustic measurements of ten snowcoaches and six snowmobiles at the southern entrance of Yellowstone National Park from the 26th through the 28th of February 2008. The measurement site location is indicated in Figure 2.

These measurements were made with three primary objectives in mind:
1) Help determine what sound testing protocols should be used to determine if *snowcoaches* meet the Best Available Technology (BAT) with respect to noise emissions.
2) Determine which *snowcoaches* meet BAT standards with respect to noise emissions.
3) Determine if there was a significant difference between *snowmobile* sound levels when tested using two different methodologies.

The measurement site was an open section of snow packed road at the south entrance of Yellowstone National Park, at the same location as was used in October 2002 for snowcoach measurements. There was a 2 to 3 foot buildup of snow in the measurement area adjacent to the road which was not ideal, however, analysis of the data indicated that this snow berm did not substantially influence the measurements.

Three microphones were setup along a line perpendicular to the road. Two were set 50 feet from the center of the over-snow vehicle travel path, one 4 feet above the snow and a second 15 feet above the snow. One microphone was set 200 feet from the center of the travel path, 4 feet above the snow. Sound levels were measured as the over-snow vehicle traveled along the roadway.

The snowcoaches tested are indicated in Table 1. Testing of the snowcoaches was guided by specifications given in SAE J1161 (Ref. 11). On the first day of testing, vehicles were measured at idle, 15 mph and 30 mph, however, due to degraded road conditions, only idle and 15 mph measurements were made on the second day. Results from these measurements are shown in Figure 1.

Table 1. Snowcoach Vehicle Description

Snowcoaches	Manufacturer	Model	Track Type*	Engine Year	Engine Size	Fuel Type
Alpen Guide	Bombardier	B-12	Skis/Tracks	2002	5.3 L V8	Gas
Goosewing Diesel Van	2006	Full Size Van, 4x4	Mattracks	2006	6.0L V8	Diesel
Rocky Mt. SC Tour	Ford	Econoline, 4x4	Mattracks	1999	6.8 L V10	Gas
Yellowstone Expeditions	Dodge	B350 Van, 2x4	Snowbuster	1994	318 V8	Gas
Yellowstone SC Tours	Ford	4x4	Mattracks	2002	V10	Gas
Goosewing Excursion	Ford	Excursion, 4x4	Mattracks	2002	6.0 L V8	Gas
Xanterra 165	Chevy	Passenger Van	Snowbuster	2001	5.7 L V8 GMC	Gas
Xanterra 431	Chevy	Express Van	Mattracks	2004	6.0 L V8	Gas
Xanterra 707	Bombardier	B-12	Skis/Tracks	1990	5.7 L V8 GMC	Gas
NPS Snowcoach	International	New Yellow Bus	Grip Trax	2003	6.0 L V8	Diesel

* Track types for the snowcoaches are as follows: Skis / Tracks – the rear wheels have been replaced with tracks while the front wheels have been replaced with skis; Mattracks – both the front and rear wheels have been replaced with triangular tracks manufactured by Mattracks; Snowbuster – are a specific manufacturer of Skis / Tracks, and Grip Trax – another specific type of triangular tracks to replace the front and rear wheels.

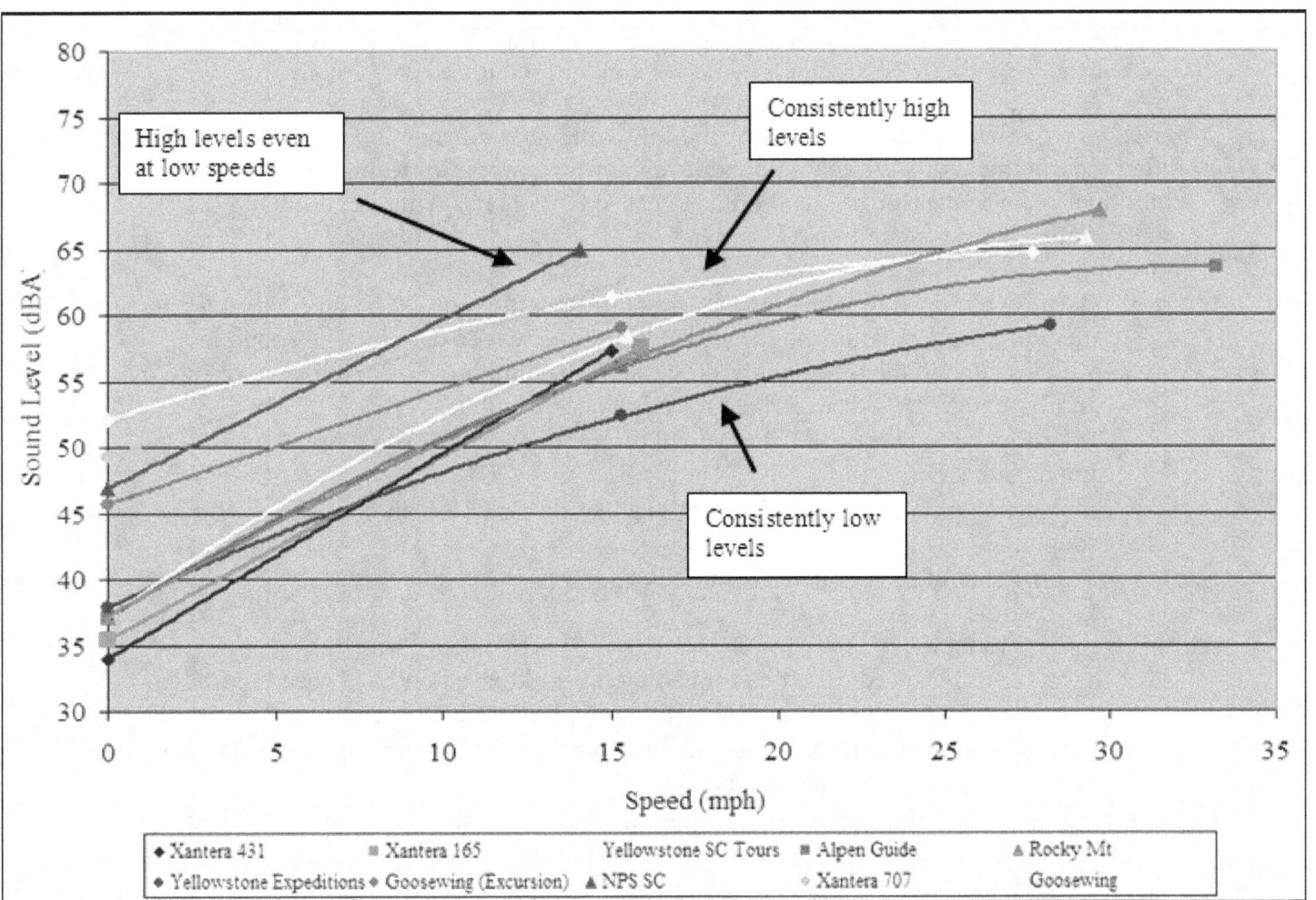

Figure 1: Sound Level vs. Speed

Snowmobiles tested are shown in Table 2. These vehicles were evaluated in order to determine if two different revisions of SAE J192 (Ref. 12) would produce different sound level results. The 1985 revision required snowmobiles to start from rest and then travel along the road at full throttle while measuring the sound level using a fast time response. The 2003 revision required snowmobiles to approach the measurement zone at 15 mph and then travel along the road at full throttle while measuring the sound level using a slow time response. On average, the 1985 revision produced results about 2 dB greater than the 2003 revision.

Table 2. Snowmobile Vehicle Description

Snowmobiles	Manufacturer	Model	Track Type*	Engine Year	Engine Size	Fuel Type
Arctic Cat TZ1	Arctic Cat	TZ1	standard	prototype	3	Gas
Arctic Cat T660 (1)	Arctic Cat	T660	standard	2004	3	Gas
Arctic Cat T660 (2)	Arctic Cat	T660	standard	2004	3	Gas
Arctic Cat T660 (3)	Arctic Cat	T660	standard	2004	3	Gas
Arctic Cat T660 (4)	Arctic Cat	T660	standard	2008	3	Gas
Arctic Cat T660 (5)	Arctic Cat	T660	standard	2006	3	Gas
Arctic Bearcat	Arctic Cat	Bearcat	wide	N/A	3	Gas

* Track types for the snowmobiles are as follows: Standard – indicates a standard width belt track with skis in the front for steering; Wide – indicates a wide belt track with skis in the front for steering.

Based on experiences during this study, the following recommendations are suggested for future measurement of snowcoach sound levels for the purpose of testing BAT conformance. The measurements should adhere to SAE J1161 with the following modifications and considerations:

- Because of the altitude, barometric pressure specifications should be expanded to include typical pressures in the parks during the winter season. The sound level variation due to the lower barometric pressure could be corrected in a manner similar to the methods described in References 5 and 6.
- If a snow berm is present, all practical efforts to remove it should be implemented.
- If a snow berm greater than 3 feet tall cannot be removed, another site should be sought.
- Testing should be conducted for three conditions
    - Idle
    - 15 mph
    - A high speed to be determined by the park based on local speed limits, e.g., 30 mph, road speed limit, or a typical cruising speed.
- A road groomer should be kept on hand in order to ensure that the road conditions do not deteriorate over the course of the testing.
- If vehicles fail to meet BAT requirements at the high speed, consideration should be given to restrictions which would still allow the snowcoach to operate in the parks, but at a reduced speed.

1. Introduction

Sounds associated with Oversnow Vehicles (OSVs), such as snowmobiles and snowcoaches, are an important management concern at Yellowstone and Grand Teton National Parks. The John A. Volpe National Transportation Systems Center's Environmental Measurement and Modeling Division (Volpe Center) is supporting the National Park Service (NPS) with implementation of the Winter Use Planning program (Ref. 7, 8, 9, 10) and supporting National Environmental Policy Act (NEPA) documents, including the 2007 Winter Use Planning / Environmental Impact Statement. As part of this support, the Volpe Center, in cooperation with the NPS, performed acoustic measurements of ten snowcoaches and six snowmobiles at the southern entrance of Yellowstone National Park from the 26th through the 28th of February 2008. The measurement site location is indicated in Figure 2.

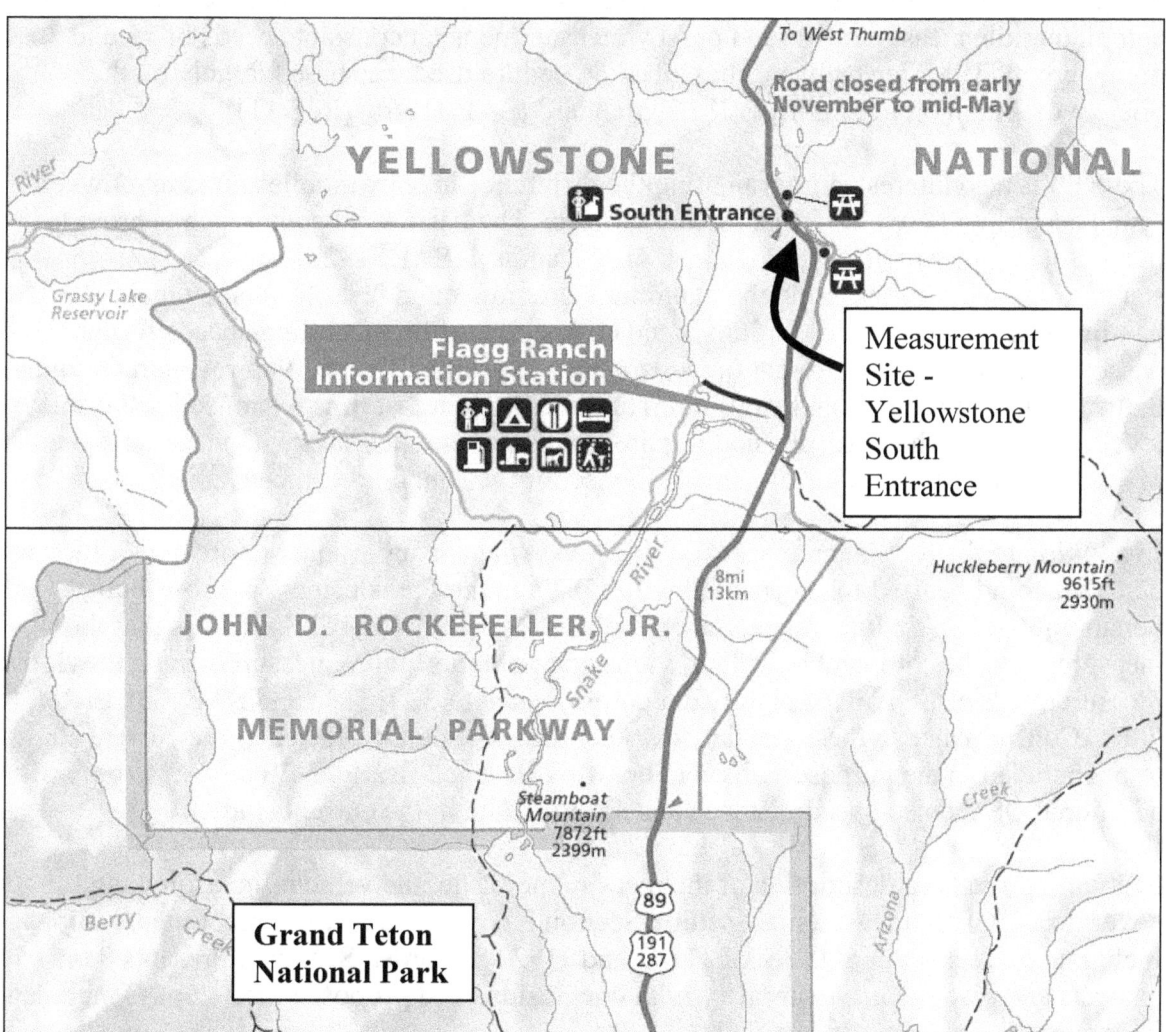

Figure 2. Location of Measurements in the John D. Rockefeller Jr. Memorial Parkway

These measurements were made with three primary objectives: 1) Help determine what sound testing protocols should be used to determine if *snowcoaches* meet the Best Available Technology (BAT) with respect to noise emissions, 2) Determine which *snowcoaches* currently operating in the parks met BAT standards with respect to noise emissions, and 3) Determine if there was a significant difference between *snowmobile* sound levels when tested using two different methodologies.

The use of BAT OSVs is one management approach to reducing the sound levels in the parks due to OSVs. The NPS wishes to determine the procedures to evaluate compliance with the snowcoach BAT requirements[*]. The winter use regulations (Federal Register, Vol. 72, No. 239, December 13, 2007, pages 70781-70804) states that, "Beginning in the 2011-2012 season, all snowcoaches must meet a sound emission requirement of no greater than 73 dBA. The Superintendent will establish procedures for determining compliance." (36 CFR 7.13l(4)(v)). In order to determine which snowcoaches are BAT compliant, idle measurements and passby measurements at constant speeds of 15 and 30 mph were made for ten snowcoaches. The procedure used was based largely on the Society of Automotive Engineers J1161 standard (SAE J1161) (Ref. 11).

The NPS is also interested in evaluating the difference in snowmobile sound level when tested according to two different methodologies. The basic testing approach is provided in the Society of Automotive Engineers J192 standard (SAE J192) (Ref. 12). This standard is used for measuring the maximum exterior sound level from snowmobile passby events. Two versions of this standard specify different starting speeds for the passby event as well as different time responses. Specifically, the 1985 revision requires that snowmobiles start from rest (0 mph) and the use of a fast time response for the sound level meter while the 2003 revision requires that snowmobiles start with an initial speed of 15 mph and the use of a slow time response for the sound level meter.

In addition to the NPS's three primary objectives, the agency wanted to obtain additional data that could be used for the modeling of OSVs in the Parks using a purpose-built version of the FAA's Integrated Noise Model (INM) (Ref. 13). This requires that the measurements be comparable with previous and potential future measurements. Previous measurements also involved significant conformance to SAE J192 and SAE J1161 (Ref. 7, 8). Conformance to these standards would allow the data collected in the current study to be included in the purpose-built version of INM. Specifically, NPS desires to obtain additional OSV noise-speed-distance and spectral data in the current study.

Section 2 provides a description of the measurement site, the vehicles measured, and cover details of the measurement setup. Section 3 covers the measurement methodology, including a description of the SAE J1161 and J192 standards. Section 4 presents data analysis and discussion. A summary with conclusions and proposed next steps is included in Section 5.

---

[*] BAT classification also includes considerations for reduced / cleaner exhaust emissions.

2. Measurement Setup and Data Collection

This section presents information on the layout and conditions of the measurement site, the vehicles measured, and the measurement equipment used.

2.1. Measurement Site Layout

The site chosen by the NPS is at the southern boundary of Yellowstone National Park. An aerial photo of the site is shown in Figure 3. The ground during the measurements was covered by at least four feet of snow. The river, shown in Figure 3, was in a ravine with a depth of approximately 25 feet. Flowing water could be faintly heard during the quietest times, however, it could not be heard during measurements. To the east of the road, there was a pull-off where tourists would stop to take photographs. The western pull-off was not present during the measurements due to snow cover. To the north there were several park buildings which did not interfere with the measurements.

Microphones were setup to the west of the road in the region free of trees. A sketch of the site showing the aerial view and profile is provided in Figure 4. Two microphones were located 50 feet from the center of the travel lane, one at 4 feet above the snow cover and one at 15 feet above the snow cover. One microphone was located 200 feet from the center of the travel lane, 4 feet above the snow cover. A meteorological system was set up 100 feet from the center of the travel lane in line with the microphones, 4 feet above the snow cover. A single measurement table was positioned 65 feet off to the side of the microphone line and all data recording equipment was setup at this location. The location of equipment is indicated in Figure 4A.

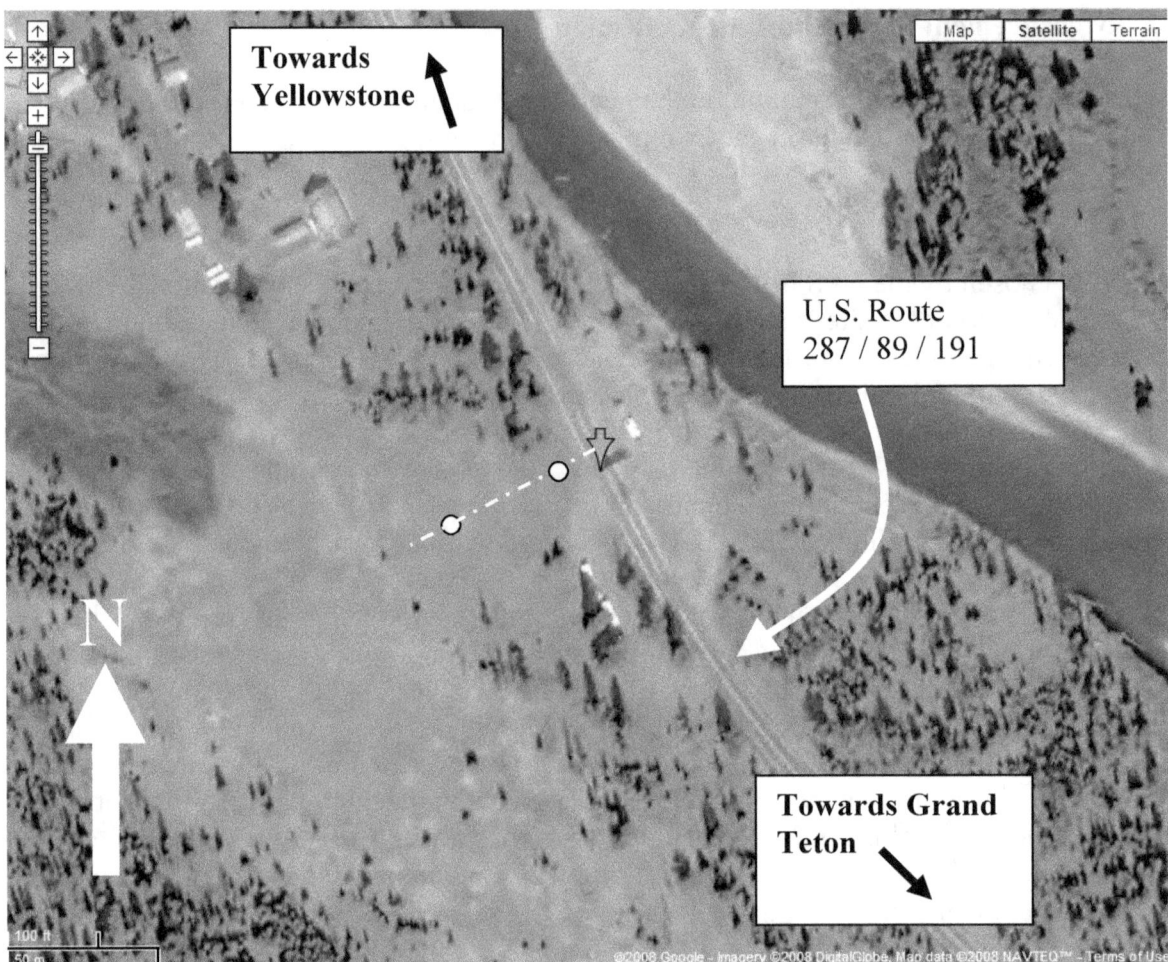

Figure 3. Aerial Photo of Measurement Site (44.133, -110.66503)[*]. White dots indicate approximate location of microphones.

Over the course of the winter, snow on the road was groomed and packed while snow off the road was not managed; thus, the snow at the microphone locations was approximately 2 to 3 feet higher than the snow along the travel lane at the measurement site. This resulted in portions of the OSV noise sources being at least partially occluded along a line of sight (LOS) to the microphone. An example of this LOS blockage can be seen in Figure 5 where the OSV tracks are occluded by the snow cover[†]. This occlusion deviates from the guidance in the SAE standards. Unfortunately, this could not be remedied. Because of this deviation, data from this study may not be directly comparable with data from other measurements, however, analysis concluded that the barrier effect was small. The barrier effect is discussed further in Section 4.

---

[*] Google Maps, 25 March 08. (http://maps.google.com/maps?f=q&hl=en&geocode=&q=44.133,+-110.66503&ie=UTF8&ll=44.132942,-110.665069&spn=0.002861,0.004989&t=h&z=18)
[†] The LOS is slightly different at the microphone because it is closer but also lower than the camera location, however, this photograph does illustrate the issue.

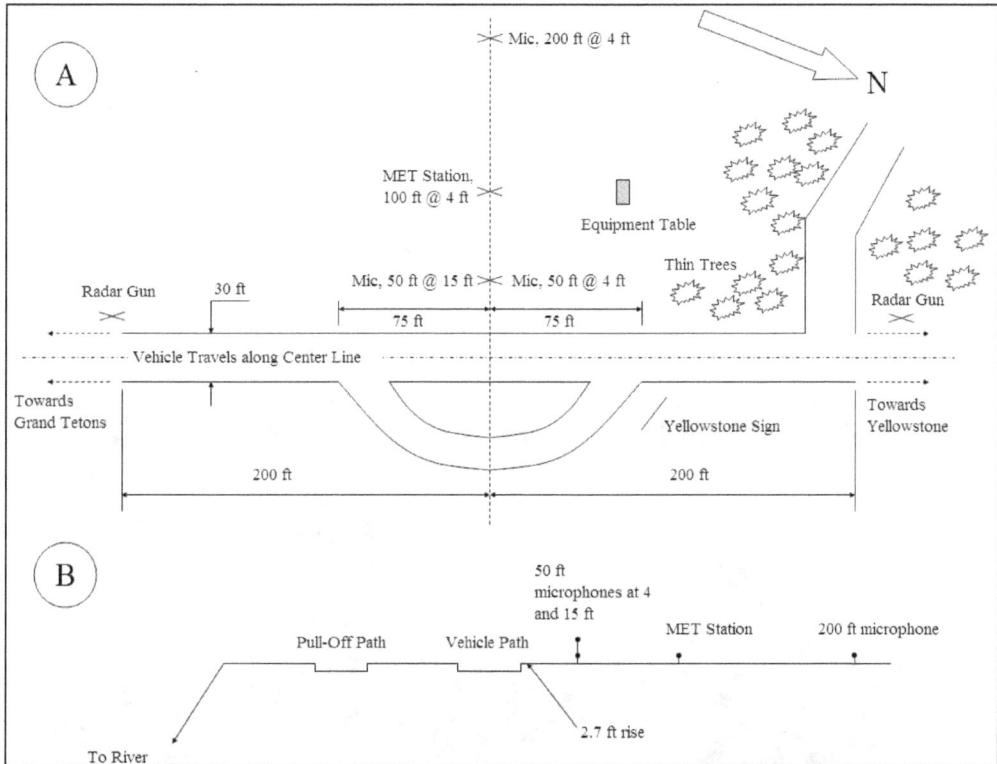

Figure 4. Measurement Site Sketch – (A) Aerial View, (B) Profile

Aside from the difference in snow height between the road and the microphone locations, the only other obstructions were the presence of some evergreen trees on the north end of the measurement area. These trees were lightly distributed and when listening to passby events, they were not observed to affect the levels at the 50 foot microphone. It is possible that they did affect portions of the data measured at the 200 foot microphone location. For this reason, only maximum levels, which were not affected by the trees, are examined for the 200 foot microphone.

Figure 5. Measurement of Yellowstone Snow Coach Tours OSV Showing Line-of-Sight Blockage of Tracks due to Accumulated Snow

2.2. Measurement Conditions

The road was not closed during the measurements so extraneous road traffic did occur, primarily during a two-hour morning interval and a two-hour afternoon interval. Extraneous traffic predominantly consisted of guided tour groups, either on snowmobile or in snowcoaches that entered Yellowstone in the morning and departed in the afternoon. In most cases, the tours stopped at the pull-off east of the measurements to take photographs. This necessitated a pause during the measurements whenever a group came through. Temperatures during the measurements ranged from 16 to 38 degrees Fahrenheit. Relative humidity ranged from 51 to 89 percent. Wind speeds ranged from 0.0 to 13.4 mph, although acoustic measurements made during wind speeds greater than 12 mph were discarded.

2.3. Vehicle Description

Over the course of the three-day study, the team measured ten snowcoaches at two speeds* and at idle for both sides of each vehicle. In addition, the team measured six snowmobiles at full throttle for two different starting conditions for both sides of each vehicle and three snowmobiles were measured at two constant (cruise) speeds for both sides of each vehicle. A summary of the vehicles' manufacturer / model type, track type, and engine characteristics is provided in Table 3 for snowcoaches and in Table 4 for snowmobiles. These tables also indicate which test types were conducted for each vehicle. Snowmobiles were from the NPS administrative fleet, with the exception of the TZ1, which was a prototype on loan to the park. Photos of the vehicles are shown in Figure 6 through Figure 15.

Table 3. Snowcoach Vehicle Description

Snowcoaches	Manufacturer	Model	Track Type*	Engine Year	Engine Size	Fuel Type	Test Type
Alpen Guide	Bombardier	B-12	Skis/Tracks	2002	5.3 L V8	Gas	cruise, idle
Goosewing Diesel Van	2006	Full Size Van, 4x4	Mattracks	2006	6.0L V8	Diesel	cruise, idle
Rocky Mt. SC Tour	Ford	Econoline, 4x4	Mattracks	1999	6.8 L V10	Gas	cruise, idle
Yellowstone Expeditions	Dodge	B350 Van, 2x4	Snowbuster	1994	318 V8	Gas	cruise, idle
Yellowstone SC Tours	Ford	4x4	Mattracks	2002	V10	Gas	cruise, idle
Goosewing Excursion	Ford	Excursion, 4x4	Mattracks	2002	6.0 L V8	Gas	cruise, idle
Xanterra 165	Chevy	Passenger Van	Snowbuster	2001	5.7 L V8 GMC	Gas	cruise, idle
Xanterra 431	Chevy	Express Van	Mattracks	2004	6.0 L V8	Gas	cruise, idle
Xanterra 707	Bombardier	B-12	Skis/Tracks	1990	5.7 L V8 GMC	Gas	cruise, idle
NPS Snowcoach	International	New Yellow Bus	Grip Trax	2003	6.0 L V8	Diesel	cruise, idle

* Track types for the snowcoaches are as follows: Skis / Tracks – the rear wheels have been replaced with tracks while the front wheels have been replaced with skis; Mattracks – both the front and rear wheels have been replaced with triangular tracks manufactured by Mattracks; Snowbuster – are a specific manufacturer of Skis / Tracks, and Grip Trax – another specific type of triangular tracks to replace the front and rear wheels.

Table 4. Snowmobile Vehicle Description

Snowmobiles	Manufacturer	Model	Track Type*	Engine Year	Engine Size	Fuel Type	Test Type
Arctic Cat TZ1	Arctic Cat	TZ1	standard	prototype	3	Gas	full throttle, cruise
Arctic Cat T660 (1)	Arctic Cat	T660	standard	2004	3	Gas	full throttle
Arctic Cat T660 (2)	Arctic Cat	T660	standard	2004	3	Gas	full throttle, cruise
Arctic Cat T660 (3)	Arctic Cat	T660	standard	2004	3	Gas	full throttle
Arctic Cat T660 (4)	Arctic Cat	T660	standard	2008	3	Gas	full throttle, cruise
Arctic Cat T660 (5)	Arctic Cat	T660	standard	2006	3	Gas	full throttle
Arctic Bearcat	Arctic Cat	Bearcat	wide	N/A	3	Gas	full throttle

* Track types for the snowmobiles are as follows: Standard – indicates a standard width belt track with skis in the front for steering; Wide – indicates a wide belt track with skis in the front for steering.

---

* Two speeds were measured when road conditions permitted, only low speeds were measured when the road conditions did not permit higher speeds.

In general, increasing the load of an OSV will increase its sound level. Two loading factors affected the test conditions, 1) the number of passengers and 2) snow conditions on the road. The snowcoaches tested had carrying capacities which ranged from eight to twenty passengers. It was not practical to test all snowcoaches at full capacity. For consistency, no passengers were used for the measurement runs. Thus, the sound levels measured represent the lowest levels that these vehicles would produce with respect to passenger load. Although the absolute sound levels during loaded operations will be higher, unloaded comparisons are reasonable to use to evaluate best available technology since they do allow for valid relative comparisons between vehicles.

Softer packed snow provides less traction and more drag, requiring OSVs to operate at higher engine speeds to obtain the same vehicle speed. It was originally planned to have a road groomer available to groom the road at least every night if not more periodically, however, it was not possible to get the groomer to maintain the road as required. Therefore, the snow pack got progressively softer with each successive day. With increasing use, the softer packed snow also developed dips and ruts which made operation of the OSVs at high speeds unsafe. Thus some snowcoaches were not tested at high speed (as represented by N/A on Table 7).

Figure 6. Purpose Built Bombardier B-12 with Skis and Tracks (Alpen Guide)

Figure 7. Customized Dodge B350 Van with "Snowbuster" Skis and Tracks (Yellowstone Expeditions)

Figure 8. Customized Econoline 4 x 4 with "Mattracks" Tracks (Rocky Mountain SC Tours)

Figure 9. Customized Van with "Mattracks" Tracks (Yellowstone SC Tours)

Figure 10. Customized Van with "Mattracks" Tracks (Xanterra 431)

Figure 11. Purpose Built Bombardier B-12 with Skis and Tracks (Xanterra 707)

Figure 12. Purpose built International Bus with "Grip Trax" Tracks (NPS Snowcoach)

Figure 13: Arctic Cat TZ1

Figure 14. Arctic Cat T660 Snowmobile

Figure 15: Artic Cat Bearcat

2.4. Equipment Description (Acoustics, Meteorology, and Speed)

Three acoustic systems, one meteorological system, and two radar guns were used to collect the requisite data.

2.4.1. Acoustic System

Each acoustic system consisted of a Larson Davis Model 824 Real Time Analyzer (LD824) a G.R.A.S. 40AE microphone and a Sony D100 DAT recorder as backup to the LD824. Figure 16 shows a block diagram of the acoustic system. The LD-LD cables were used to connect the microphone / preamp to the LD824 positioned at the measurement table, so that all acoustic systems could be monitored at a single location. Care was taken to insure that no connections were exposed to moisture. This involved using plastic bags to cover connections and, wherever possible, keeping cables off the snow.

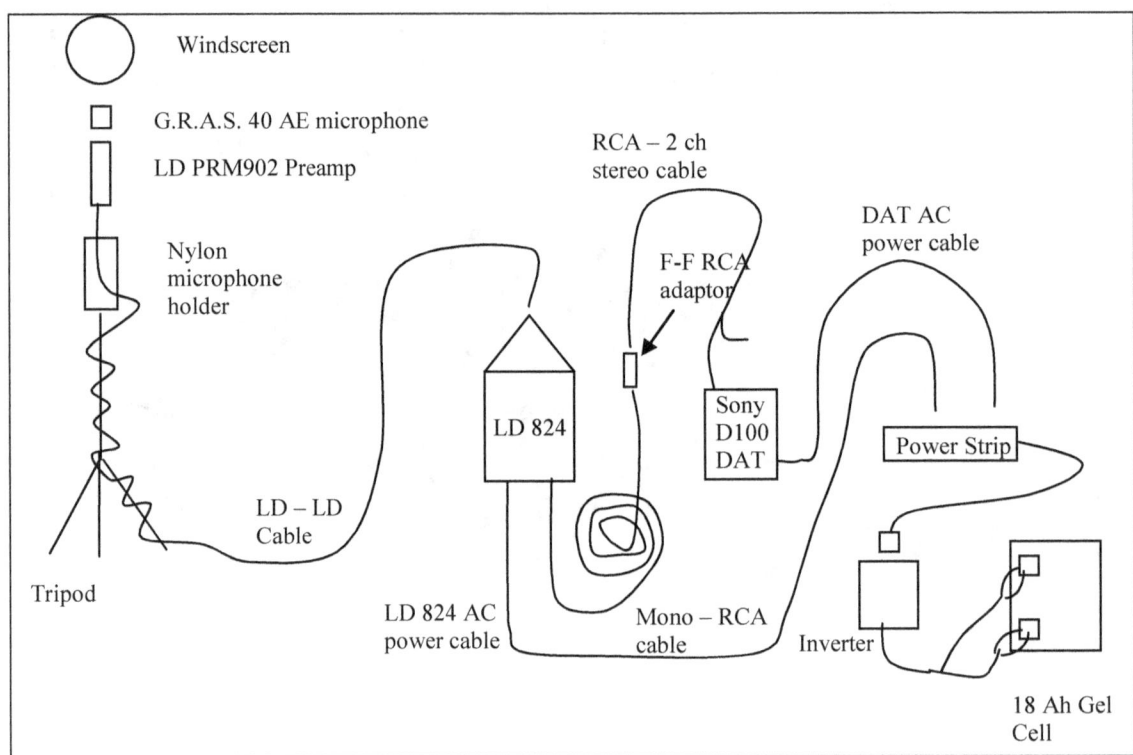

Figure 16. Acoustic System Setup

One-half second samples were used for the 50 foot microphone at a height of 4 feet above the snow cover while 1 second samples were used for the remaining microphones[*]. At all microphone locations the overall L_{Aeq} and L_{ASmx} were stored in the LD824s. One-third octave band levels were also stored for each sample. Additional information on the LD824 and the Sony D100 DAT is provided in Appendix A and Appendix B.

2.4.2. Meteorological Equipment

A Qualimetrics Transportable Automated Meteorological Station (TAMS) was used to measure wind speed, wind direction, relative humidity, and air temperature at one-second intervals. A complete TAMS system consists of a sensor unit and a control / display unit that displays real time meteorological data. These battery-powered stations are portable and well suited for remote sampling. Table 5 shows the metrics collected and the operating accuracy and Figure 17 shows a block diagram of the system. The unit's sensors were placed on a tripod at a height of 4 feet above the snow cover. Data from the control unit is automatically displayed and saved onto a laptop using TAMS Analysis software which can be viewed in real-time. Additional information on the TAMS unit is provided in Appendix C.

---

[*] A 1 second sample was used for the 50 foot microphone at a height of 4 feet above the snow cover on the first day. For the second and third day the sample period was changed to ½ second.

Table 5. TAMS Measurement Capabilities

Metric	Range	Resolution	Accuracy
Wind Speed	2 to 55 mph	1 mph	1 mph or 5% of range
Wind direction	360°	10	root mean standard error of 18°
Temperature	-9 to 110 °F	1 °F	1 °F
Relative Humidity	0 to 100 %	1%	3%

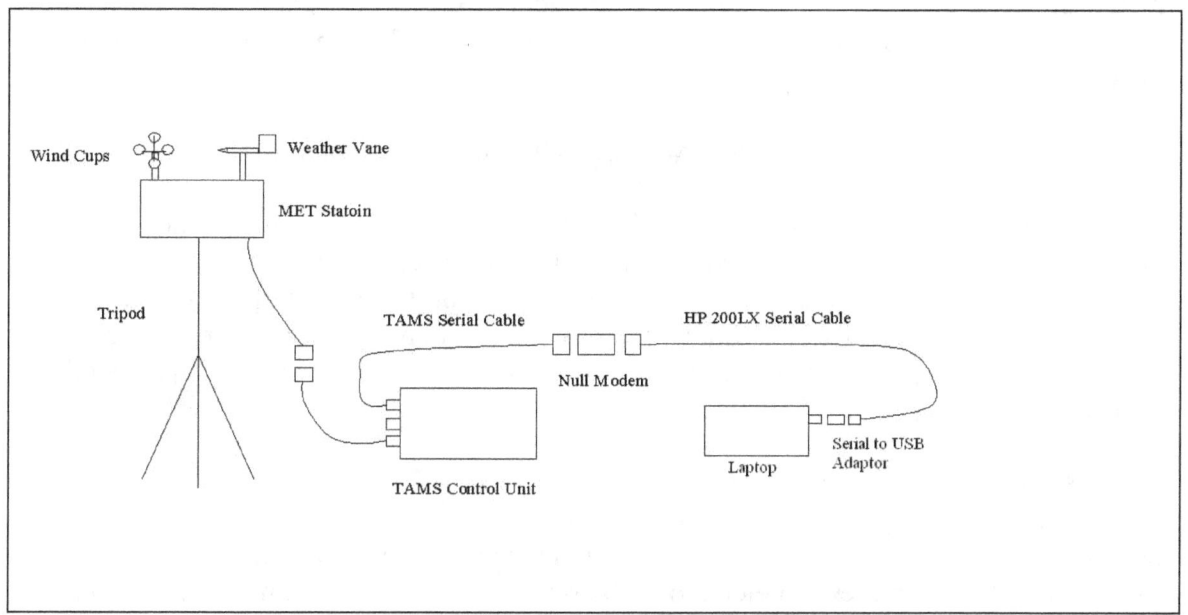

Figure 17. TAMS Setup

2.4.3. Vehicle Speed Collection (Doppler-Radar Gun)

Two Doppler-radar detectors were used to measure vehicle speed. One was situated at each end of the 400-foot span along the travel path centered at the microphone line. Their locations are indicated in Figure 4. Each radar detector was used to measure the speed of vehicles approaching the given radar detector location.

3. Measurement Methodology

This section details the measurement methodologies with specific references to SAE J1161 and J192. The measurements were conducted jointly by the NPS and Volpe. The NPS provided the vehicles, the measurement site, tools to help OSV drivers control their speed during the measurements, and staff to assist. Volpe provided acoustic, meteorological, and speed measurement equipment (see Section 2) and staff to conduct the measurements. Details of the measurement protocol and a sample log sheet are given in Appendix D.

3.1. Methodology in Accordance with SAE International Standard J1161

SAE J1161 specifies a methodology for the measurement of operational sound levels for OSVs. The measurement area is required to be an open region of packed snow at least 2 inches deep[*] which is free of reflecting surfaces. The snow pack is required to be sufficient to support the OSV without penetration with no more than 3 inches of loose snow on top of the packed snow. An illustration of the measurement area is provided in Figure 18.

The OSV approaches the measurement area at a speed of 15 mph and maintains this speed throughout the area. The microphone / sound level meter is required to be positioned 50 feet from the center of the travel lane and 4 feet above the snow cover (see Figure 18). The sound level meter should be set to measure the maximum A-weighted sound pressure level with slow time weighting (L_{ASmx}) as the vehicle passes at a constant speed of 15 mph. Measurements are repeated until three L_{ASmx} values are within 2 dB. These L_{ASmx} values are then arithmetically averaged and rounded to the nearest integer. The measurements are conducted for the vehicle traveling in both directions and the side of the vehicle with the highest average is reported as the sound level for the OSV. When drivers are traveling towards Grand Teton, the microphones are on their right hand side, this is documented as the "right side" of the vehicle. When drivers are traveling towards Yellowstone, the microphones are on their left hand side, this is documented as the "left side" of the vehicle. An example of the sound level calculation is given in Table 6.

During the measurements, the standard specifies that atmospheric temperature, pressure, relative humidity, wind speed and direction be measured at recorded intervals of not less than 1 hour. Further details are given in Reference 11.

---

[*] (alternatively grass)

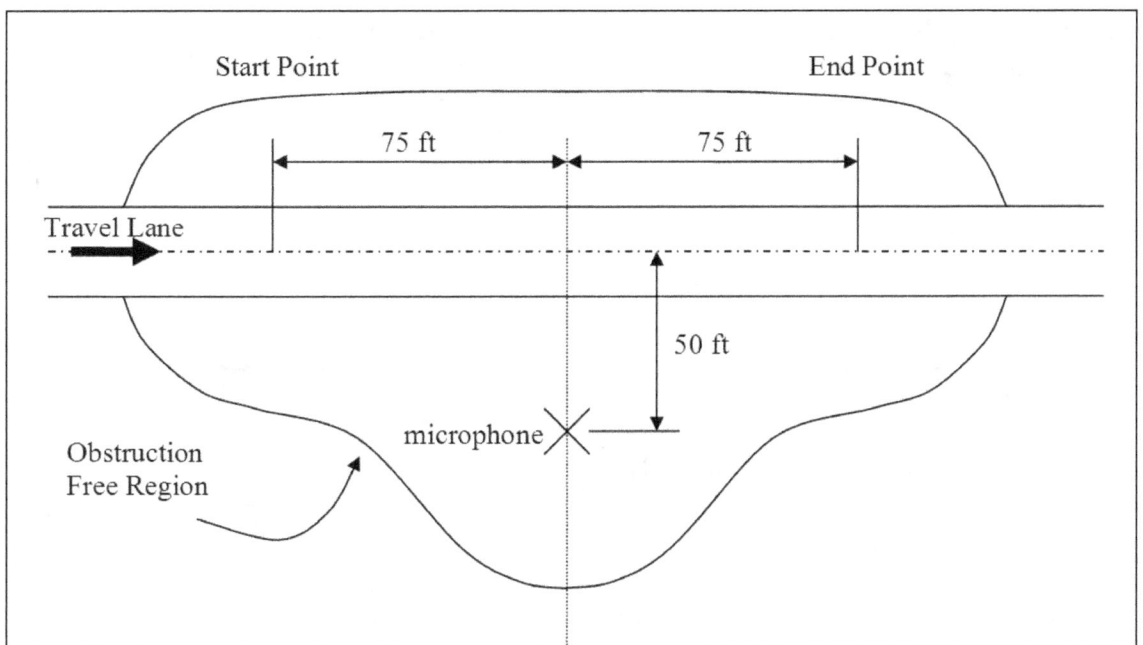

Figure 18. SAE J1161 / SAE J192 Measurement Area Layout

Table 6. Example Sound Level Calculation for SAE J1161

Passby	LASmax, dB(A)			within 2 dB?	Average	Final Level
Left Side	64	65	65	yes	65	65
Right Side	62	61	62	yes	62	

3.2. Methodology in Accordance with SAE International Standard J192

SAE J192 specifies a methodology for the measurement of maximum exterior sound levels for snowmobiles. This standard is the same as SAE J1161 with the exception that it is specified for snowmobiles only (not OSVs in general), that vehicles are intended to be measured at full throttle rather than at continuous speed, and, for the 1985 revision, a fast time response is specified.

Two revisions of SAE J192, 1985 and 2003, were used during the study. The 1985 revision specifies that the snowmobile start with zero initial speed at the start point (see Figure 18) and then maintain full throttle as it travels through the measurement area. The 2003 revision specifies that the snowmobile approach the start point with an initial speed of 15 mph and then maintain full throttle as it travels through the measurement area. Further details of this methodology are provided in Reference 12.

3.3. Additional Measurements

In order to maximize the data obtained during the study, vehicles were operated under additional operating conditions, additional microphones were utilized, and the measurements were made over an extended measurement area. The measurement area was extended to cover a region from 200 feet on either side of the microphone line instead of the 75 feet on either side*. See Figure 19. The extension to 200 feet on either side provided sufficient distance to assure a 10 dB drop from the maximum level, so that Sound Exposure Levels (SELs) could also be readily computed.

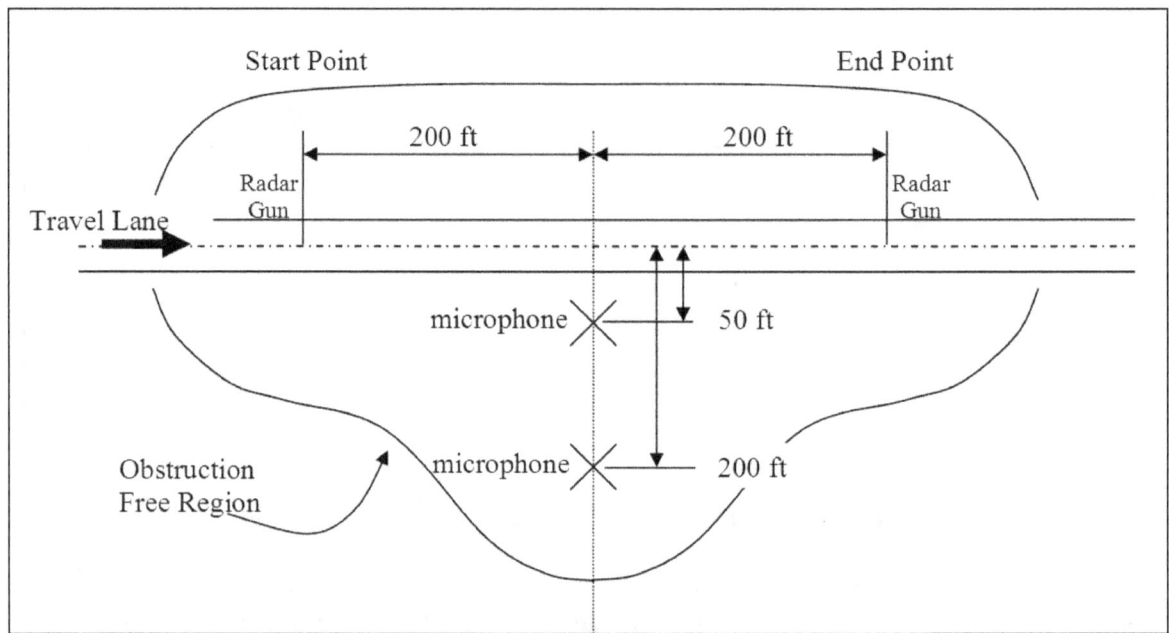

Figure 19. Extended Measurement Area Layout

In addition to the microphone required by SAE J1161 and SAE J192, two additional microphones were used in the study. One microphone was positioned at the 50 foot location but with a height of 15 feet above the snow cover. This microphone was intended to provide a measurement that was less influenced by the ground cover. A second microphone was set at 200 feet from the travel lane at a height of 4 feet above the snow cover. This microphone was intended to provide a second measurement near the ground so that comparisons can be made with the 50 foot microphone in order to evaluate the ground attenuation over distance at the site.

Additional operating conditions were also measured for the snow coaches and included measurements at idle and at high constant speed. Idle measurements consisted of one minute of L_{Aeq} measurements made for each side of the vehicle. High constant speed tests were vehicle dependent, that is, the vehicles were operated at the highest speed that could be maintained over the course, typically about 30 mph.

3.4. Meteorological Measurements

Meteorological measurements were made throughout the measurement period and recorded at 1 second intervals. This is a higher temporal resolution than required by the SAE J1161 and J192 standards, but allows for increased quality control of the measurements, especially with respect to wind speed. Temperature, relative humidity,

---

* For the SAE J192 type measurements the 150 ft measurement area was maintained because increasing the area would effect the speed reached at the microphone line and beyond.

wind speed, and wind direction were measured by using a TAMS system located 100 feet from the center of the road, inline with the microphones. These data were displayed and recorded on a laptop computer which was located at the measurement table, as discussed in Section 2.4.1. The wind speed was monitored in order to avoid making measurement runs when the wind speed was above 12 mph.

4. Results and Analysis

Processed results and analysis are presented in this section. Analysis focuses on two of the primary objectives: 1) to determine which snowcoaches had the Best Available Technology (BAT) with respect to noise emissions, and 2) to determine if there was a significant difference between snowmobile sound levels when tested using two different methodologies. Additional analysis examines the attenuation due to the snow bank and the ground attenuation over the distance between the 50 and 200 foot microphones.

4.1. Ambient and Meteorological Conditions

A summary of the meteorological data is provided in Table 7. The primary concern during the measurements was wind speed. Although wind speeds exceeded 12 mph on both the 26th and the 28th, care was taken to avoid making measurements during such periods. High wind speeds were not observed on the 27th.

Table 7. Summary of Meteorological Conditions During the Measurements

Day	Temperature (°F)			Relative Humidity (%)			Wind Speed (mph)		
	Min.	Max.	Avg.	Min.	Max.	Avg.	Min.	Max.	Avg.
26-Feb-08	26.1	36.5	29.2	54.0	78.0	69.1	0.0	12.8	4.9
27-Feb-08	16.2	38.1	26.6	51.0	89.0	68.6	0.0	5.4	1.6
28-Feb-08	24.4	28.9	26.2	78.0	88.0	84.7	0.0	13.4	4.3

During measurements, the primary sound sources which contributed to the ambient level included OSVs operated at other locations in the parks, birds, running water, and wind / vegetation sound. Ambient levels ranged from 34 to 37 dB(A) on the 26th, 19 to 27 dB(A) on the 27th, and 34 to 39 dB(A) on the 28th. The lower ambient levels observed on the 27th, for example 19 dB(A), are likely influenced by the instrumentation noise floor, thus the true ambient level may be even lower. The higher ambient levels included more human generated sound and higher wind speeds. For the purposes of this study, these estimates of the ambient level provide a conservative limit for comparison with vehicle noise measurements.

4.2. Best Available Technology (BAT) for Snowcoach Noise Emissions

Ten snowcoaches were measured following guidelines in SAE J1161. These data have been processed to provide information that can be used to evaluate BAT qualification. Note, because there was partial Line-of-Sight (LOS) blockage due to the difference in height between the road and the snow cover adjacent to the road where the measurements were made, relative comparisons can be made between the ten snowcoaches measured as part of this project, but comparisons to snowcoaches measured at other times or locations may not be appropriate (but see Section 5).

4.2.1. Sound Level Time History

Although the primary noise metric for SAE J1161 and SAE J192 is the maximum A-weighted sound pressure level with slow exponential time weighted averaging, L_{ASmx}, an additional useful metric for examining time histories is the equivalent continuous sound pressure level, $L_{Aeq\Delta t}$, where Δt is the duration of the measurement. (L_{Aeqs} are needed for the computation of the Sound Exposure Level (SEL) as will be discussed in Section 4.2.4.)

An example $L_{Aeq\Delta t}$ time history is shown in Figure 20 over 1 second intervals ($L_{Aeq1sec}$) for the Goosewing Diesel Van snowcoach measured during a low speed passby on February 26th. This figure shows the general pattern observed for all passby events: values increase as the vehicle approaches the microphone center line; there is a peak region; values decrease as the OSV travels away from the microphone center line. Another important feature to observe is that the peak level is at least 10 dB greater than the lowest approach and lowest departure levels. This indicates that a clean passby was acquired and that analysis using these data will not be unduly contaminated with sound sources other than the vehicle under study. An example time history profile is given in Figure 21 for a high-speed passby of the Goosewing Diesel Van snowcoach.

An example idle measurement time history is shown in Figure 22 for the Goosewing Diesel Van snowcoach measured on February 26th. The profile is different here because the vehicle is stationary and neither approaches nor departs the microphone line during the measurement. Because the level does not change significantly over the measurement period it is not readily apparent that this is a clean measurement. In order to verify that these data are not contaminated by extraneous sounds, these data are compared with ambient levels. The ambient level was about 17 dB(A) lower than the idle level for this particular measurement and is therefore also considered a clean measurement. An example sound level time history is provided for each vehicle at each speed measured in Appendix E.

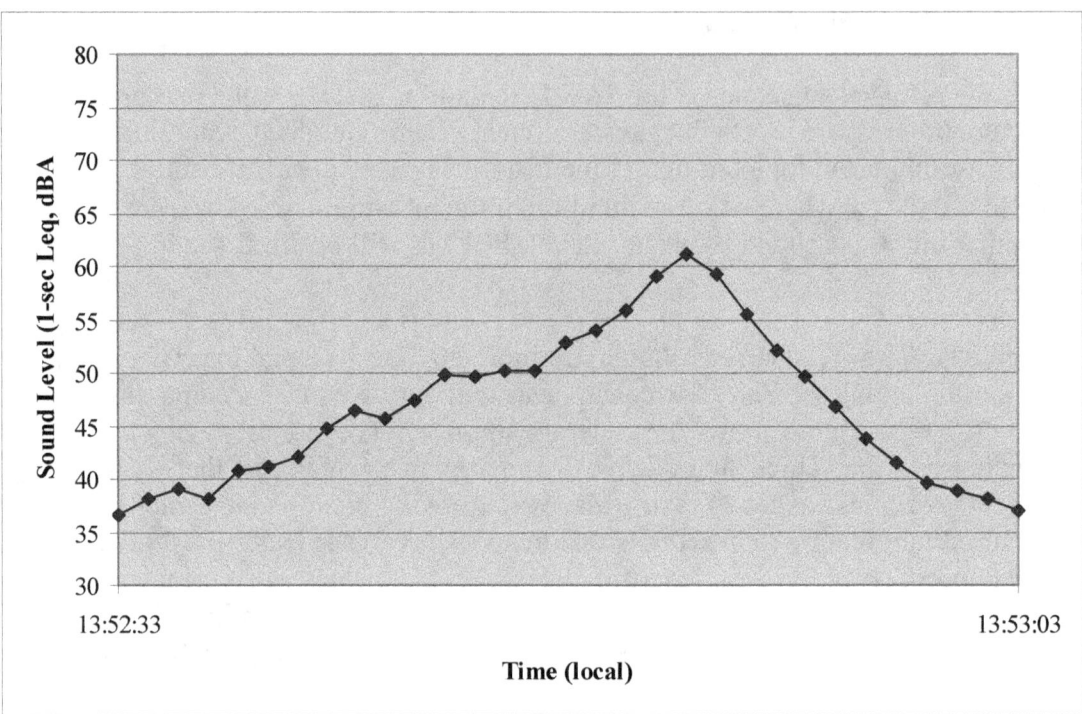

Figure 20. Goosewing Diesel Van, Right Side at Low (15 mph) Speed (Feb 26th)

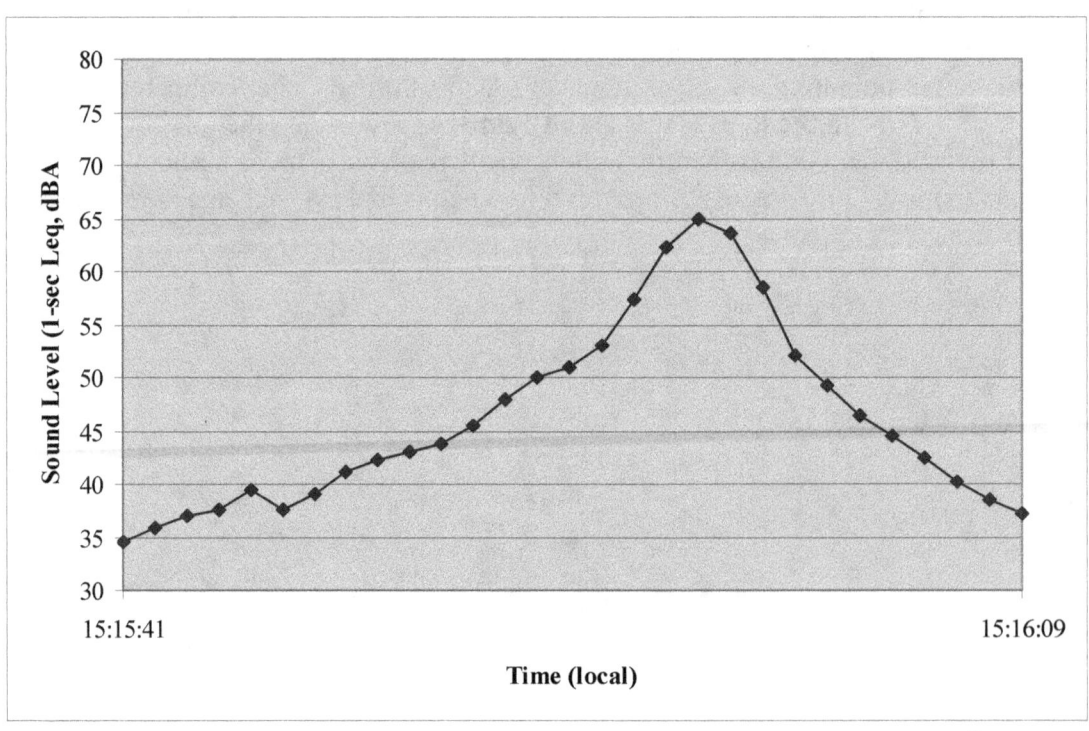

Figure 21. Goosewing Diesel Van, Left Side at High (30 mph) Speed (Feb 26th)

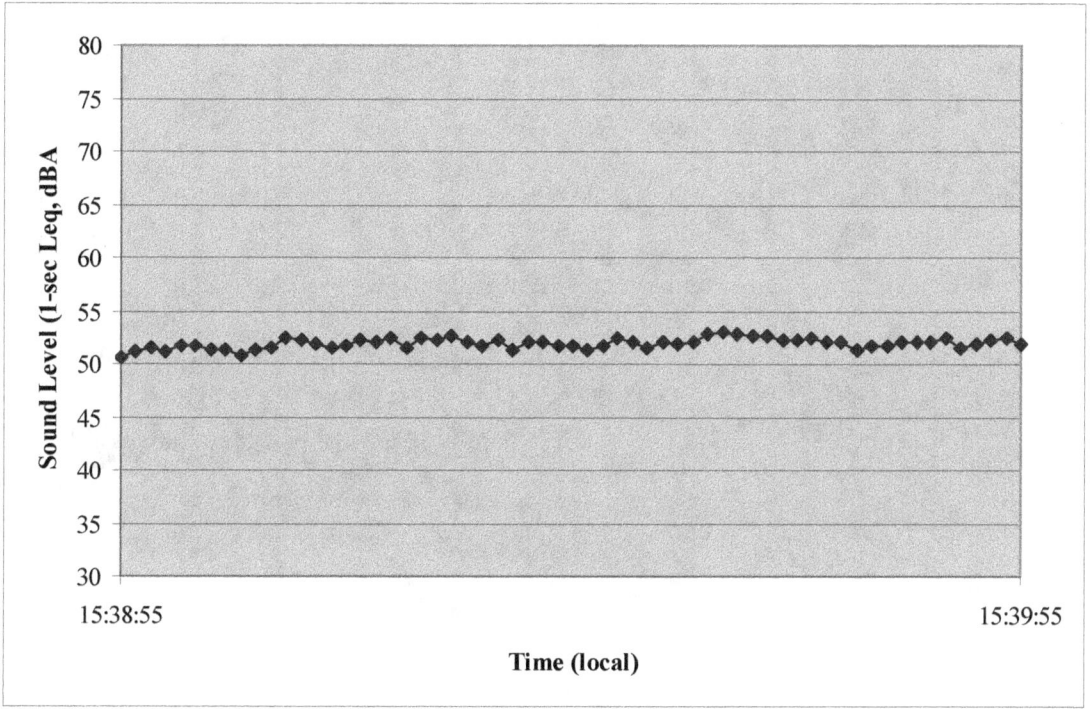

Figure 22. Goosewing Diesel Van, Left Side at Idle (Feb 26th)

An example of the difference between the sound levels on the two sides of a vehicle is given in Figure 23 for the Xanterra 707. Here, the right side of the vehicle had the highest idle level. This side of the vehicle had a strong aspirated sound which showed up as a rhythmic level variation between 2 and 5 dB per sample. The right side of the vehicle had a lower, more steady idle level. Typically, time varying sounds are more noticeable than sounds that do not fluctuate. Therefore this rhythmic level variation may be more objectionable to park visitors. If the high side could be controlled so that its noise emission was the same as the lower side, this vehicle's idle level could be reduced by 8 to 9 dB. Note, that the Xanterra 707's engine is not fully enclosed (see Figure 11), therefore efforts to enclose the engine may help reduce the idle level.

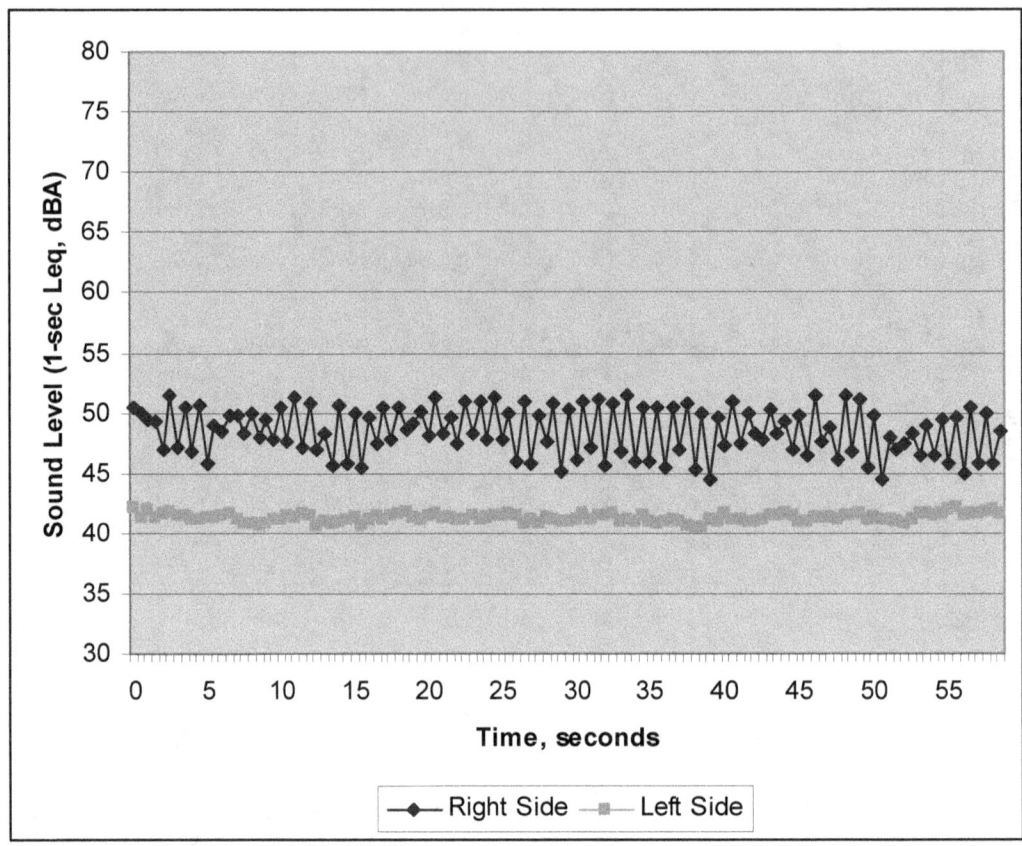

Figure 23. Xanterra 707 Idle (Feb 27th)

4.2.2. Overall Level (L_{ASmx}) Associated with Each Vehicle

Table 8 rank orders the ten snowcoaches measured from highest to lowest maximum A-weighted sound pressure levels with slow time weighting (L_{ASmx}) for a constant speed of 15 mph (nominal). The values listed in Table 8 are the arithmetically averaged L_{ASmx} values for the passby direction which yielded the highest average value. Averages of the measured speeds are specified in the last column. Table 9 rank orders the ten snowcoaches measured from highest to lowest based on the arithmetically averaged L_{ASmx} values for the passby direction which yielded the highest average value for a constant speed of 30 mph (nominal). Averages of the measured speeds are specified in the last column.

The Yellowstone Expeditions and Alpen Guide snowcoaches consistently had lower sound levels at both speeds. The NPS snowcoach had high sound levels at low speeds but high speed measurements were not possible. Even so, the NPS snowcoaches' low speed sound levels are comparable to other snowcoach high speed measurements. A summary of all the L_{ASmx} values that were used to produce the reported maximum levels is provided in Appendix F.

Table 8. L_{ASmx} for Loudest Side of Vehicle at Low Speed, dB(A)

Vehicle	Vehicle Side[#]	Average L_{ASmx}	Average Speed of Runs, mph
NPS SC	Left	64.9	14.1
Xanterra 707	Right	62.5	15.3
Goosewing Diesel Van	Right	61.3	15.0
Goosewing Excursion	Right	59.0	15.3
Yellowstone SC Tours	Right	58.4	15.5
Xanterra 165	Left	57.5	15.9
Xanterra 431	Right	57.3	15.0
Rocky Mt	Right	56.3	15.3
Alpen Guide	Right	56.0	15.3
Yellowstone Expeditions	Right	52.3	15.3

[#] "Left" indicates left side of vehicle is closest to the microphones. "Right" indicates right side of vehicle is closest to the microphones.

Table 9. L_{ASmx} for Loudest Side of Vehicle at High Speed, dB(A)

Vehicle	Vehicle Side[#]	Average L_{ASmx}	Average Speed of Runs, mph
Rocky Mt	Left	67.9	29.7
Yellowstone SC Tours	Right	65.8	29.3
Goosewing Diesel Van	Left	64.7	27.7
Alpen Guide	Right	63.6	33.2
Yellowstone Expeditions	Left	59.1	28.2
Goosewing Excursion	N/A	N/A	N/A
Xanterra 165	N/A	N/A	N/A
Xanterra 431	N/A	N/A	N/A
Xanterra 707	N/A	N/A	N/A
NPS SC	N/A	N/A	N/A

[#] "Left" indicates left side of vehicle is closest to the microphones. "Right" indicates right side of vehicle is closest to the microphones.

Table 10 rank orders the ten snowcoaches measured from highest to lowest based on the arithmetically averaged L_{Aeq} values for the idle orientation with the highest level. In this case, instead of averaging passby measurements, a one-minute period of L_{eq1sec} values were averaged. The levels at idle of several snowcoaches were close to the ambient, indicating that at least some of the energy in the reported level is contaminated by the ambient. This would cause these levels to overestimate the true idle levels of these vehicles, so these levels should be considered a conservative estimate of the idle levels. Idle measurements with levels that are within 10 dB of the ambient are marked with an asterisk (*) in Table 10.

Whereas levels during passby measurements are due to a combination of engine / exhaust, drive train, and ski / track noise, idle levels are predominantly due to engine / exhaust noise alone. Both the Xanterra 165 and Xanterra 431 had low idle levels. These were late model Chevrolet van conversions, while for example, the Xanterra 707 was an older Bombardier B-12. Both the Xanterra 165 and Xanterra 707 were fitted with 5.7 liter V8 GMC engines, so modifications to the engine alone is not always sufficient to reduce noise emissions. Modifications may also need to be made, for example, to the exhaust and shielding. Further investigation into OSV specifications is needed to determine all factors that contribute to BAT.

Table 10. L_{Aeq} for Loudest Side of Vehicle at Idle (0 mph), dB(A)

Vehicle	Vehicle Side[#]	Average L_{Aeq}
Goosewing Diesel Van	Left	51.9
Xanterra 707	Right	48.5
NPS SC	Left	46.8
Goosewing Excursion	Left	45.6
Yellowstone Expeditions	Left	37.8[*]
Rocky Mt	Right	37.2[*]
Alpen Guide	Right	37.1[*]
Yellowstone SC Tours	Left	37.0[*]
Xanterra 165	Left	35.4[*]
Xanterra 431	Right	34.0[*]

[#] "Left" indicates left side of vehicle is closest to the microphones. "Right" indicates right side of vehicle is closest to the microphones.
[*] Idle measurements with levels that are within 10 dB of the ambient level

4.2.3. Noise vs. Speed Plots

The passby levels for high and low speeds were combined with idle levels to generate noise-speed curves as shown in Figure 24. This plot essentially presents the data from Table 8, Table 9, and Table 10 graphically for easier comparison. The Yellowstone Expeditions snowcoach consistently has the lowest overall noise profile over the speed range tested while the Goosewing Diesel Van has the highest. The Xanterra 707 and the NPS snowcoach also had high levels at idle and the highest level at low speed. They did not have data at high speed. Interestingly, the Xanterra 707 and the Alpen Guide are both Bombardier B-12 models. The Alpen Guide had a 2002 5.3 liter gasoline V8 while the Xanterra 707 had a 1990 5.7 liter gasoline V8. The Xanterra 707 was consistently noisier than the Alpen Guide, especially at low speeds. The difference is most pronounced at idle, where, as mentioned previously the Xanterra 707 had an unusual rhythmic variation in sound level.

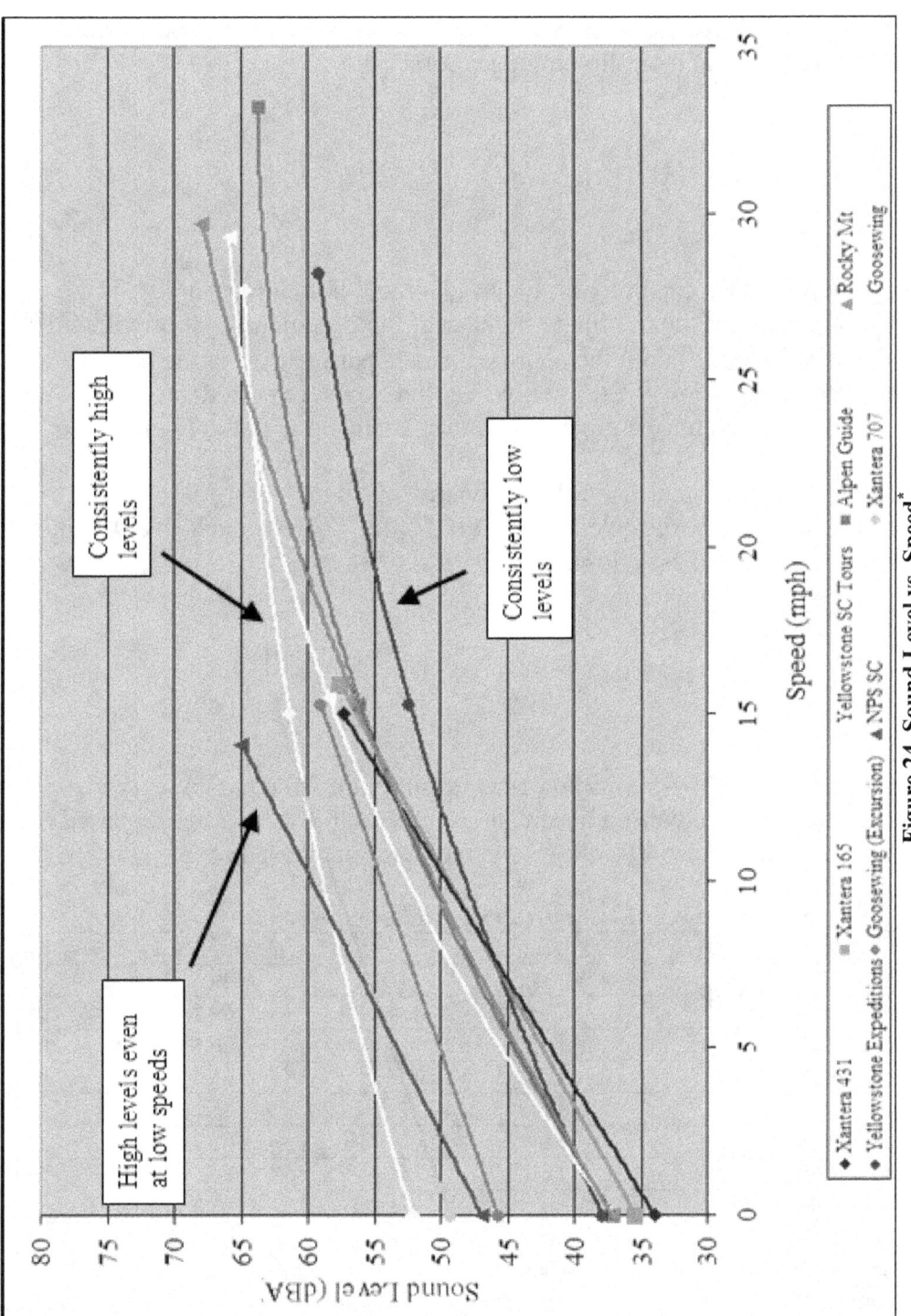

Figure 24. Sound Level vs. Speed*

---

* (Note theoretical noise-speed curves should asymptote to a constant as the speed approaches 0 mph, however, more speeds would be required in order to capture the shape of this asymptote. For modeling purposes in the parks, three points are sufficient.)

4.2.4. Sound Exposure Level (SEL), dB(A)

The overall sound exposure level (SEL) due to a snowcoach passby gives another indication of the noise performance of the snowcoach. The SEL can be estimated from the stored LD824 data by using the following equation:

$$SEL = 10\log_{10}\left(\Delta t \sum_{i=1}^{N} 10^{L_{Aeq\Delta t}(i)/10}\right),$$

where Δt is the time period for each record, $L_{Aeq\Delta t}(i)$ is the ith $L_{Aeq\Delta t}$ measured, i is the record increment that includes the minimum number of measurements that provide a 10 dB drop from the peak $L_{Aeq\Delta t}$ for both the approach and departure during the passby. Because the SEL is an integration of the energy over the exposure period, a vehicle with a slow increase / decrease can have a greater SEL than a vehicle with the highest L_{ASmx}.

The SELs reported in Table 11 and Table 12 are energy averages of all passby measurements that were used in the calculation of L_{ASmx} for Table 8 and Table 9. Thus, the SEL equation was modified as follows:

$$SEL = 10\log_{10}\left(\frac{\Delta t}{M} \sum_{i=1}^{N} 10^{L_{Aeq\Delta t}(i)/10}\right),$$

where M is the number of passby events and i iterates through three passby events. Allowing for order reversals between adjacent snowcoaches, the rank ordering based on SEL for these snowcoaches generally follows the rank ordering based on L_{ASmx}.

Table 11. Average SEL for each Event at Low Speed, dB(A)

Vehicle	Vehicle Side[#]	Average SEL
Xanterra 707	Right	69.1
Goosewing Diesel Van	Right	67.2
NPS SC	Left	66.6
Goosewing Excursion	Right	65.9
Yellowstone SC Tours	Right	65.2
Xanterra 431	Left	65.0
Xanterra 165	Left	64.7
Alpen Guide	Right	63.8
Rocky Mt	Right	63.3
Yellowstone Expeditions	Right	59.1

[#] "Left" indicates left side of vehicle is closest to the microphones. "Right" indicates right side of vehicle is closest to the microphones.

Table 12. Average SEL for each Event at High Speed, dB(A)

Vehicle	Vehicle Side[#]	Average SEL
Rocky Mt	Left	72.7
Yellowstone SC Tours	Right	70.5
Goosewing Diesel Van	Left	70.1
Alpen Guide	Right	68.4
Yellowstone Expeditions	Left	64.0
Goosewing Excursion	N/A	N/A
Xanterra 165	N/A	N/A
Xanterra 431	N/A	N/A
Xanterra 707	N/A	N/A
NPS SC	N/A	N/A

[#] "Left" indicates left side of vehicle is closest to the microphones. "Right" indicates right side of vehicle is closest to the microphones.

4.2.5. Sound Level Spectra

For each record in the sound level time histories, there is an associated unweighted one-third octave band spectrum. The spectrum associated with the L_{ASmx}[*] during one passby of the Goosewing Diesel Van measured at low speed on February 26th is shown as an example in Figure 25. Here, the general pattern of decreasing level with increasing one-third octave band center frequency can be seen. Additionally, occasional increases in a third-octave band can be seen. Generally these spectral peaks are associated with engine or drive train harmonics at discrete frequencies and can be expected to increase in frequency with increasing engine speed. The effect of increasing engine speed on spectral peaks or tones can be seen in Figure 26 and Figure 27. In Figure 26 a spectral peak can be seen in the 80 Hz one-third octave band for the Yellowstone Expedition snowcoach traveling at slow speed. When this snowcoach travels at high speed (with increased engine rpm), the spectral peak shifts to the 125 Hz one-third octave band.

Unweighted one-third octave band spectra associated with the passby time histories in Appendix E are provided in Appendix G.

---

[*] The $L_{Aeq\Delta t}$ is most appropriate for examining time histories, but the L_{ASmx} gives a better indication of the maximum level during a passby.

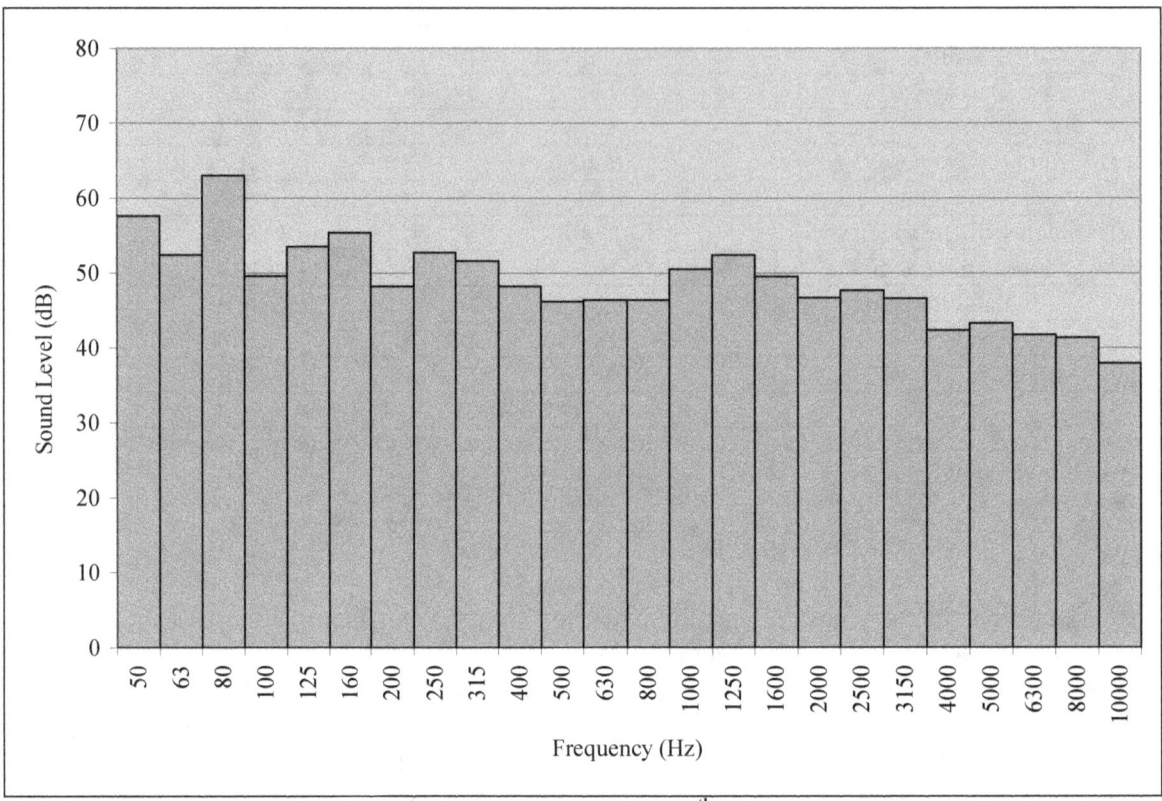

Figure 25. Goosewing Diesel Van, Right Side (Feb 26th, 13:52) Spectra for Low Speed

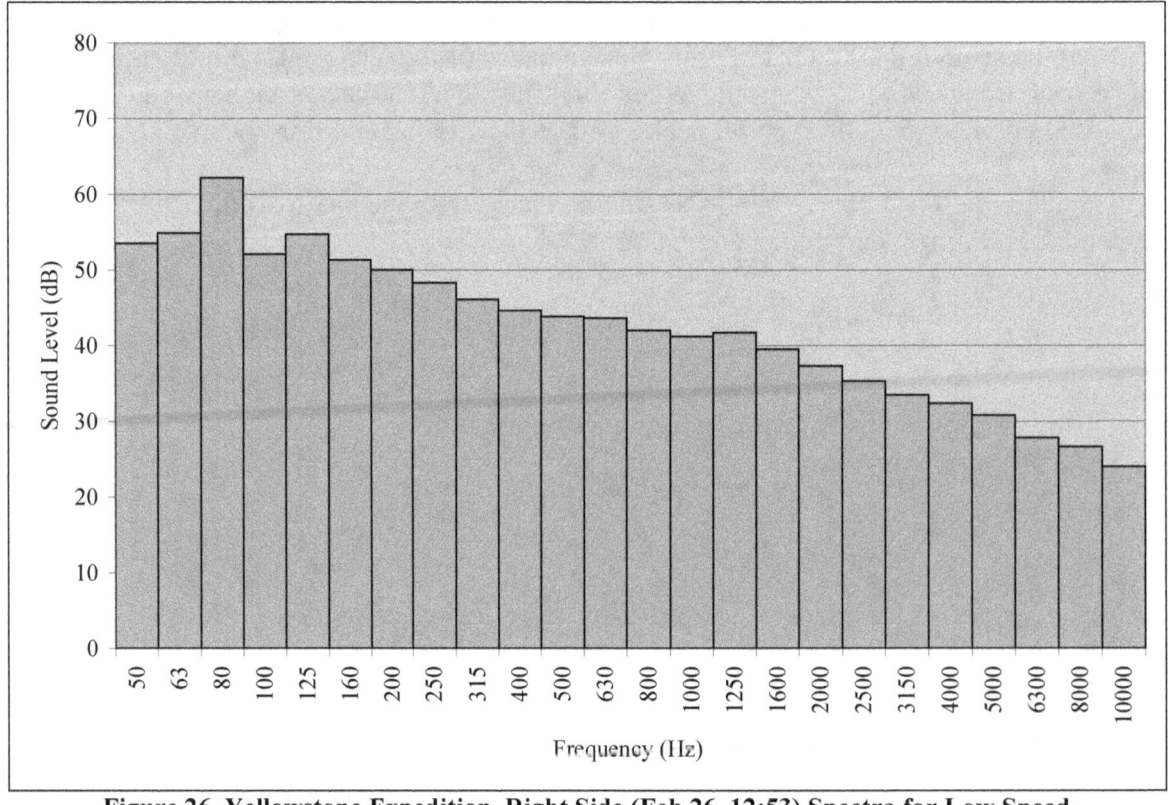

Figure 26. Yellowstone Expedition, Right Side (Feb 26, 12:53) Spectra for Low Speed

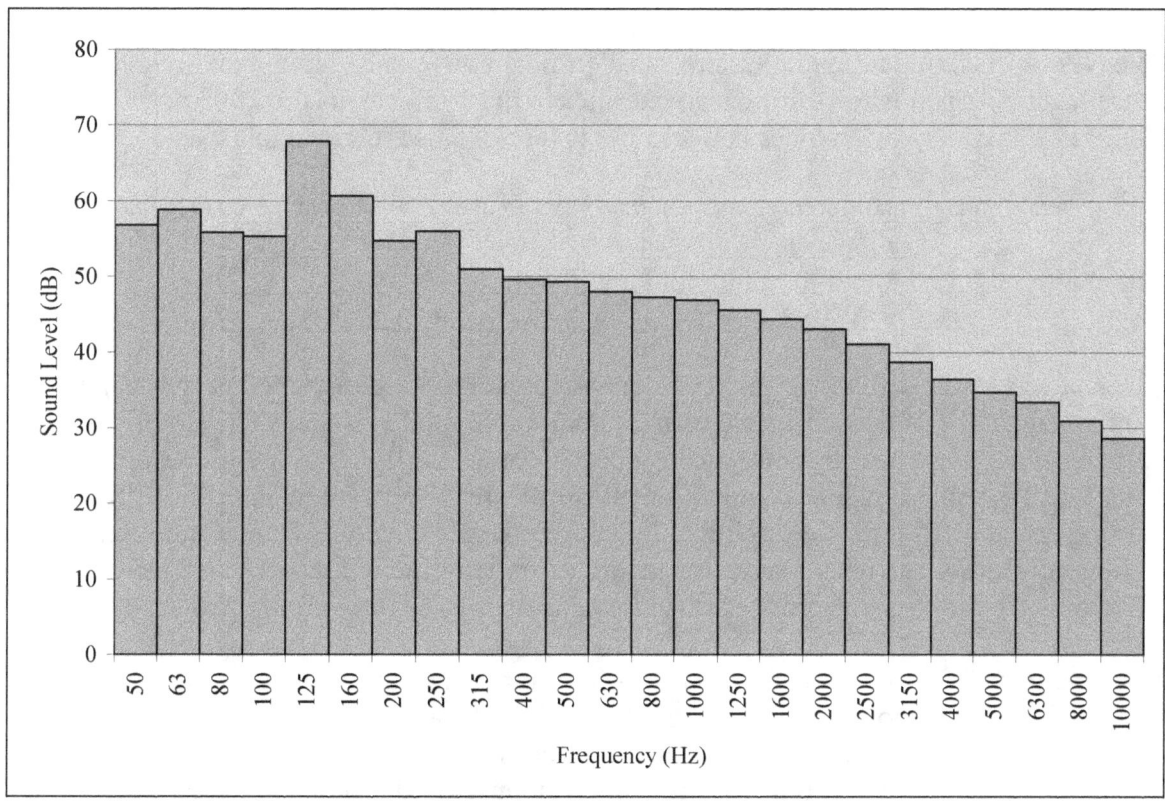

Figure 27. Yellowstone Expedition, Left Side (Feb 26, 14:18) Spectra for High Speed

4.3. Comparison of SAE J192 (1985) and SAE J192 (2003)

The NPS wished to determine if there is a significant difference in the maximum exterior noise level obtained when testing snowmobiles to the 1985 versus the 2003 revision of SAE J192. The key differences between the two revisions are given in Table 13.

Table 13: Differences between the 1985 and 2003 Revisions of SAE J192

Revision	Start Speed, mph	Time Response
1985	0	fast
2003	15	slow

The analysis of the difference between the two revisions was conducted in two stages. In the first stage, the effect of starting from either 0 or 15 mph was analyzed. In this stage, slow response was used for both starting speeds. In general, changing the starting speed from 0 to 15 mph produces essentially identical results. On average, the results from a starting speed of 15 mph were about 0.2 dB higher than the results from a starting speed of 0 mph. Run-to-run results have a standard deviation of about 0.4 dB, so these variations are on the same order of magnitude as the run-to-run variations.

In the next stage of this analysis, the levels for 0 mph starting speeds were determined using a fast time response, L_{AFmax}. Since the different starting speeds were shown to produce essentially the same results, the differences between the two methods can be mostly attributed to the difference in the time response used. Thus, the effect of the different time response can be seen by comparing the 1985 and 2003 revisions as given in Table 14. The difference between the two revisions is shown in the last column, where the results for the 2003 revision are subtracted from the results from the 1985 revision. Here it can be seen that using the fast time response results in values that are a little less than 2 dB higher than those for the slow time response[*]. This is expected since the fast time response allows the level to rise faster as the snowmobile passes by the microphone. Therefore, vehicles which are just under the acceptable sound limit based on the 2003 revision may turn out to be over the limit when using the 1985 revision.

---

[*] While this difference is greater than the run-to-run variation, it is less than the standard's stated measurement uncertainty of +/- 2 dB. (Ref. 12 comment 6.7)

Table 14: Comparison

Vehicle	0 - Full Vehicle Side[#]	L_{AFmax} (1985 rev, 0 mph start)	15 - Full Vehicle Side[#]	L_{ASmax} (2003 rev, 15 mph start)	Difference (15 mph - 0 mph test)
TZ1	Right	71.4	Left	69.7	1.7
T660(2)	Right	65.5	Right	63.6	1.9
T660(3)	Left	64.4	Left	63.5	0.9
T660(4)	Right	66.9	Right	64.9	2.0
T660(5)	Right	65.7	Right	64.1	1.6
Bear Cat	Left	74.1	Right	71.8	2.3
				Average Difference	1.7

[#] "Left" indicates left side of vehicle is closest to the microphones. "Right" indicates right side of vehicle is closest to the microphones.

4.4. Snowmobiles at Constant Speed

During the study, there was an opportunity to measure three of the snowmobiles, TZ1, T660(2), and T660(4), at constant speeds, much in the same way as was done for the snowcoaches. The results of these opportunistic measurements are given in the following tables. The rank order was preserved for both high and low speeds for both the L_{ASmx} and for the SEL. The TZ1 consistently had the highest levels.

Table 15. L_{ASmx} for Loudest Side of Vehicle at Low Speed, dB(A)

Vehicle	Vehicle Side[#]	Average L_{ASmx}	Average Speed of Runs, mph
TZ1	Left	63.8	24.5
T660(2)	Right	61.5	25.5
T660(4)	Left	60.0	26.0

[#] "Left" indicates left side of vehicle is closest to the microphones. "Right" indicates right side of vehicle is closest to the microphones.

Table 16. L_{ASmx} for Loudest Side of Vehicle at High Speed, dB(A)

Vehicle	Vehicle Side[#]	Average L_{ASmx}	Average Speed of Runs, mph
TZ1	Left	67.1	45.5
T660(2)	Right	65.8	43.0
T660(4)	Left	63.6	43.0

[#] "Left" indicates left side of vehicle is closest to the microphones. "Right" indicates right side of vehicle is closest to the microphones.

Table 17. Average SEL for each Event at Low Speed, dB(A)

Vehicle	Vehicle Side[#]	Average SEL	Average Speed of Runs, mph
TZ1	Left	72.2	24.5
T660(2)	Right	69.5	25.5
T660(4)	Left	68.6	26.0

[#] "Left" indicates left side of vehicle is closest to the microphones. "Right" indicates right side of vehicle is closest to the microphones.

Table 18. Average SEL for each Event at High Speed, dB(A)

Vehicle	Vehicle Side[#]	Average SEL	Average Speed of Runs, mph
TZ1	Left	74.2	45.5
T660(2)	Right	73.1	43.0
T660(4)	Left	70.7	43.0

[#] "Left" indicates left side of vehicle is closest to the microphones. "Right" indicates right side of vehicle is closest to the microphones.

4.5. Ground Attenuation

In order to estimate the attenuation of snow-covered ground for the OSVs, the maximum L_{ASmx} values for passby events at the 50 foot location 4 feet above the snow cover were compared with the L_{ASmx} values at the 200 foot location 4 feet above the snow cover. Attenuation due to spherical divergence for a point source is $20 \times \log_{10}(2)$ per doubling of distance, or about 6 dB per doubling. So, the attenuation due to divergence between the 50 and 200 foot microphones should be 12 dB. Atmospheric attenuation is small over this distance for these sounds, and can effectively be neglected. Once spherical divergence is accounted for, the differences between the levels at the 50 foot and 200 foot microphones can effectively be attributed to excess ground attenuation. The estimated ground attenuation for three snowcoaches are shown in Table 19. These snowcoaches were selected from the measurements on the 27th, the day which had the lowest overall winds. On average, the ground attenuation was about 4.7 dB. Ground attenuation is typically limited to about 20 dB as the distance between the source and receiver increases, due to the effects of turbulence and scattering (Ref. 14).

To see how this ground attenuation compared with ground attenuation for other surfaces, the Federal Highway Administration's (FHWA's) Traffic Noise Model (TNM) (Ref. 15, 16, 17) was used to evaluate the ground attenuation of a typical lawn and of field grass. The ground attenuation was determined over the range from 50 to 200 feet from the source at a height of 4 feet above the ground using medium trucks. On average over this distance, lawn had a ground attenuation of 2.5 dB per distance doubling, field grass had a ground attenuation of 3.0 dB per distance doubling, granular snow had a ground attenuation of 3.2 dB per distance doubling and powder snow had a ground attenuation of 3.6 dB per distance doubling. It is reasonable to obtain a ground attenuation of 4.7 dB per distance doubling for the deep snow present at the measurement site.

Table 19. Ground Effect Results for Three Vehicles Measured on February 27th.

Vehicle	Vehicle Side#	Vehicle Speed (mph)	Wind Speed (mph)	50 ft. LASmx (dBA)	200 ft. LASmx (dBA)	LASmx Difference (50 ft - 200ft) (dBA)	Spherical Divergence (dB)	Remaining Attenuation (dB)	Ground Effects per Doubling (dB)
Goose wing Ex	Right	15.0	0.0	59.1	37.5	21.6	12.0	9.6	4.8
Goose wing Ex	Left	15.5	2.0	58.1	37.3	20.8	12.0	8.8	4.4
Goose wing Ex	Right	15.0	0.0	58.8	38.0	20.8	12.0	8.8	4.4
Goose wing Ex	Left	15.0	0.0	57.3	36.0	21.3	12.0	9.3	4.7
Goose wing Ex	Right	16.0	2.2	59.1	37.9	21.2	12.0	9.2	4.6
Goose wing Ex	Left	16.0	2.7	58.1	37.4	20.7	12.0	8.7	4.4
Goose wing Ex	Right	28.0	0.0	65.4	44.6	20.8	12.0	8.8	4.4
Goose wing Ex	Left	28.0	2.5	64.3	42.7	21.6	12.0	9.6	4.8
Xanterra 165	Right	16.0	0.0	56.8	33.4	23.4	12.0	11.4	5.7
Xanterra 165	Right	16.0	0.0	56.2	34.6	21.6	12.0	9.6	4.8
Xanterra 165	Right	24.0	0.0	60.9	38.9	22.0	12.0	10.0	5.0
Xanterra 431	Right	15.0	0.0	58.1	38.1	20.0	12.0	8.0	4.0
Xanterra 431	Right	15.0	0.0	56.0	34.4	21.6	12.0	9.6	4.8
Xanterra 431	Left	15.0	0.0	56.4	35.3	21.1	12.0	9.1	4.6
Xanterra 431	Right	15.0	0.0	57.7	36.0	21.7	12.0	9.7	4.9
Xanterra 431	Left	15.0	2.7	57.9	35.0	22.9	12.0	10.9	5.5

Avg Ground Effects = 4.7 dB per Doubling

"Left" indicates left side of vehicle is closest to the microphones. "Right" indicates right side of vehicle is closest to the microphones.

4.6. Snow Bank (Barrier) Effects

As mentioned in Section 2.1, the measurement site was less than ideal with regard to variations in the vertical profile. To investigate the influence of the wayside snow bank on the measured sound levels, one-third octave band spectra from both 50 foot microphones were compared. This approach seemed appropriate since the 15 foot microphone had clear line-of-sight to all OSV sound sources under study. This was not the case for the 4 foot microphones due to the snow bank. Sample spectra are shown in Figure 28 for a Xanterra 165 passby event at low speed. In this figure, the 15 foot one-third octave bands are shown next to the corresponding 4 foot one-third octave bands. It can be seen that for all but the highest frequencies the two spectral profiles are about the same, that is, the difference does not seem to be frequency dependent. If the snow bank were significantly shielding the 4 foot microphone, its spectrum would roll off at about 6 dB per octave relative to the 15 foot microphone at higher frequencies. Since this roll-off is not observed, it seems that the barrier effect of the wayside snow bank is likely relatively small for the 4 foot microphone. This is especially true for the overall level.

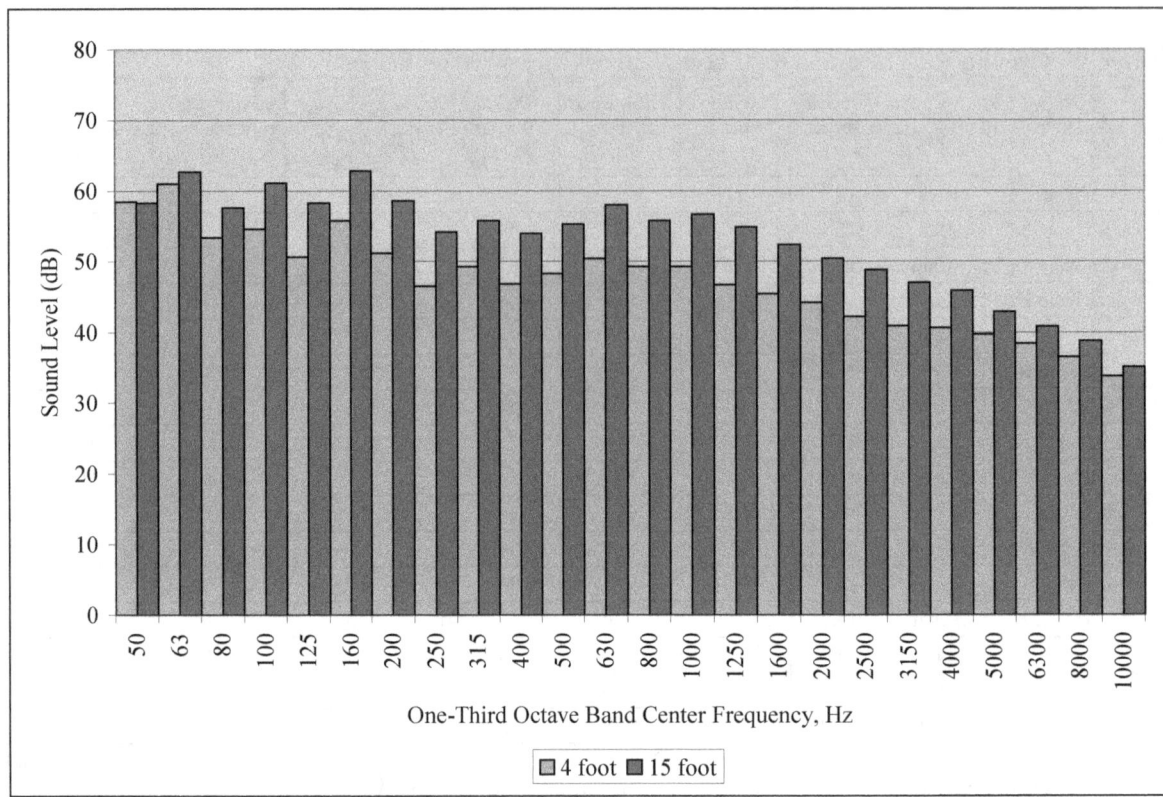

Figure 28. Xanterra 165, Left Side (Feb 27, 11:50) One-Third Octave Band Spectra for the Microphones 4 and 15 foot above the Snow Cover at 50 feet from the Travel Lane

5. Summary and Conclusions

The maximum A-weighted sound pressure levels with slow time response were measured for ten snowcoaches during constant speed passby events. Measurements were made for each vehicle side and the side with the highest average (based on three measurements within 2 dB) was assigned as the vehicle's passby noise level. Additionally, idle noise levels were also determined. Based on the low speed results, the NPS snowcoach had the highest noise levels while the Yellowstone Expeditions snowcoach had the lowest noise levels. High speed results were incomplete due to deteriorating road conditions on the second and third measurement days, however, of the five snowcoaches that were measured at high speed, the Rocky Mountain snowcoach had the highest level and the Yellowstone Expeditions had the lowest level. The Xanterra 165 and the Xanterra 431 had the lowest measured idle levels. More complete rankings can be found in Table 8, Table 9, and Table 10. All snow coaches met NPS' 73 dB BAT sound level limit for the speeds tested. Two Bombardier B-12 snowcoaches, one operated by Alpen Guide and one operated by Xanterra had very different sound levels. The Xanterra B-12 was about 11 dB higher than the Alpen Guide at idle and about 7 dB higher at 15 mph. Design differences between these two snowcoaches may provide insight into methods to decrease the Xanterra B-12 sound levels. The Yellowstone Expeditions' snowcoach provides one example of a snowcoach which may meet new limits should the NPS decide to lower the limit in the future.

The maximum A-weighted sound pressure levels with fast and slow time response were measured for six snowmobiles during full throttle passby events for two starting speeds 1) 0 mph and 2) 15 mph. Multiple measurements were made for each vehicle side and the side with the highest average (based on three measurements within 2 dB) was assigned as the vehicle's passby noise level. The two starting speeds produced similar results with an average difference of 0.2 dB(A). The use of the fast time response resulted in an increase of about 2 dB on average over the slow time response. The 1985 revision of SAE J192 specifies a fast time response and a starting speed of 0 mph. The 2003 revision specifies a slow time response and a starting speed of 15 mph. Therefore, snowmobiles tested according to the 1985 revision are expected to produce levels about 2 dB higher than the 2003 revision, due solely to the difference in specified time response.

Although the height difference between the road and the ground on either side of the road was initially a concern, analysis of the spectra between the 4 foot high and 15 foot high microphones suggests that the snow-barrier effect did not substantially effect the overall results. Therefore the data measured on these three days may be cautiously compared with other measurements made in accordance with SAE J1161 and SAE J192.

It was also observed that snow conditions on the road deteriorated substantially over the course of testing which made it impossible to test at high speeds on the second day. The 15 foot microphone was useful as a quality check, however, provided that the site is properly maintained, future testing should not require this microphone. Based on experiences during this study, the following recommendations are suggested for future measurement of snowcoach sound levels for the purpose of testing BAT conformance.

The measurements should adhere to SAE J1161 with the following modifications and considerations:

- Because of the altitude, barometric pressure specifications should be expanded to include typical pressures in the parks during the winter season. The sound level variation due to the lower barometric pressure could be corrected in a manner similar to the methods described in References 5 and 6.
- If a snow berm is present, all practical efforts to remove it should be implemented.
- If a snow berm greater than 3 feet tall cannot be removed, another site should be sought.
- Testing should be conducted for three conditions
    - Idle
    - 15 mph
    - A high speed to be determined by the park based on local speed limits, e.g., 30 mph, road speed limit, or a typical cruising speed.
- A road groomer should be kept on hand in order to ensure that the road conditions do not deteriorate over the course of the testing.
- If vehicles fail to meet BAT requirements at the high speed, consideration should be given to restrictions which would still allow the snowcoach to operate in the parks, but at a reduced speed.

Appendix A: Larson Davis Model 824 Sound Level Meter

This appendix provides some useful information on the settings, setup, and use of Larson Davis Model 824 sound level meters.

A.1 LD824 Settings

One-half second samples were used for the 50 foot microphone at a height of 4 feet above the snow cover while 1 second samples were used for the remaining microphones[*]. The following broadband metrics were stored with the same time periods as the captured data:

- Average A-weighted sound level (L_{Aeq}),
- Maximum A-weighted slow and fast values (L_{ASmax}, L_{AFmax}),
- Minimum A-weighted slow and fast values (L_{ASmin}, L_{AFmin}).

Table 20. Larson Davis 824 real time analyzer settings.

	Broadband	Narrow Band
Detector	Slow	Slow
Weighting	A	Flat
Bandwidth	N/A	1/3
Period (seconds)	½ or 1	½ or 1
Resolution (dB)	0.1	0.1

A.2 Check/apply Setup Parameters

1. Press **SETUP** button to access 824 virtual instruments menu.
2. Highlight **Edit Settings: *SLM&RTA SSA>** using ↑↓ keys, then select by pressing √ key.
3. On the Settings Page, use ↑↓ keys to highlight desired feature, then √, to access parameters (refer to summary of settings on the following page). Features include: **SLM, RTA, Intervals, Time History, Ln, Triggering, Advanced.** *Note: Gain settings can be found in SLM feature.*
4. Within each feature, use ↑↓ to toggle through parameters. Select parameter using √.
5. Parameter settings will appear in drop down menus. Use ↑↓ to highlight, then √ to apply desired settings. Back out of menu directories using ← key.
6. Repeat steps 3 through 5 for each feature until all desired settings have been applied.

---

[*] A 1 second sample was used for the 50 foot microphone at a height of 4 feet above the snow cover on the first day. For the second and third day the sample period was changed to ½ second.

A.3 Setting the Clock
1. Press **TOOLS** button. Use ↑↓ to highlight Clock/Timer. Press √ to select.
2. Press √ again to enter **Clock/Timer** menu. Use ↑↓ to highlight **Current Time** and √ to select.
3. Use ← → to select hours, minutes, seconds, and ↑↓ to alter values.
4. Press √ to set time. Time will set and increment from the moment √ is pressed.
5. Use same process to set **Current Date** and **Day Of Week**.
6. To view clock in real-time, from any feature, press **TOOLS**. Use ↑↓ to highlight **Clock/Timer**. Press √ to select.

A.4 Calibration
1. Press **TOOLS** button. Use ↑↓ to highlight **Calibration**. Press √ to select.
2. Press √ again to enter **Calibration** menu. Ensure that **Cal Level** is set to desired level. If not, select (√) **Cal Level**; use arrow keys to change value, and √ to enter appropriate setting.
3. Use ↑↓ to highlight **Change,** then √. You will be prompted to make sure calibrator is active. If so, select (√) **YES**. Wait while calibration offset is being performed. Screen will go to Calibration readout with date, time and offset.
4. Calibration is complete. You may now select your desired **VIEW**.

A.5 Collect and Monitor Data
1. Press **VIEW**. Use ↑↓ to highlight preferred instrument (SLM or RTA), then select (√).
2. Press **RUN/STOP** to collect data. Press again to stop data collection.

Appendix B: Sony Model TCD-100 DAT Recorder

This appendix provides useful information on the setup and use of Sony Model TCD-100 DAT recorders.

B.1 Sony DAT Settings

Sampling Frequency	Input	Microphone Attenuation	Input Range
44.1 kHz	Mic	0 dB	-6 dB @ 94 dB, 1 kHz

B.2 Check / Apply Setup Parameters
Sampling rate: 48 kHz or 44.1 kHz (CD compatible), 32 kHz in "LP" (half speed) mode
Input mode: Always set to "LINE"
Gain control: Always set to "MANUAL"

Tape duration: A 60-m (197-ft) tape will run for 2 hours at normal speed; 4 hours at LP half speed. Although the "LP" mode does not use linear PCM encoding, testing has established that amplitude linearity is good to within +/- 1.5 dB down to 85 dB below full-scale (0 VU).

B.3 Setting the Clock
1. Hold **CLOCK/SET** button until the date display flashes.
2. Adjust the value of flashing digits with the + & - buttons. Press **CLOCK/SET** button to advance to next digit. When setting seconds, the value can only be set to zero.
3. Start DAT clock by pressing **CLOCK/SET** button when time-base seconds reach zero. Internal clock ticks at own rate. Toggle between 12-hour time and 24-hour time by pressing and holding the + button for 3 seconds while the time is displayed.
4. To view clock in real-time, from any feature, press **TOOLS**. Use ↑↓ to highlight **Clock/Timer**. Press √ to select.

B.4 Operation Notes

1. The Input level control has a friction-lock, but care should be taken to prevent movement of the control after calibration. A small section of stiff card (or a piece special 3M tape) can be used to secure the control knob. This single control knob affects both channels equally. No independent adjustment of input levels is possible.
2. "AVLS" and Volume +/- controls affect the headphone output, but not the LINE output.
3. Allow ~30 seconds when recording, for ID marker to be written. Note that "START-ID" flashes in the display while the marker is being written. If recording is stopped before writing of marker is complete, the ID table may be scrambled.
4. Before recording, set display to "Absolute Time" by pressing the **COUNTER** button until "A-TIME" appears in the display. This indicates the absolute time from the start of the tape. Note that in "LP" mode, the value displayed is half of the actual elapsed time.
5. 6. To open the deck, slide the **OPEN** switch. Wait for "OPEN" to appear in the display before attempting to physically open the unit. Note that the deck will not open if the **LOCK** switch is engaged.
6. The **LOCK** switch will prevent most controls from operating - an important exception is the recording level control.
7. To view input signal levels, press the **RECORD** button. To prepare for recording, press **RECORD** and **PAUSE**. To record immediately, press **RECORD** and **PLAY**.
8. If the pause-record mode is enabled for more than 5 minute, the unit will time-out and drop out of this mode.
9. During recording, the current sample rate can be displayed by pressing the **PLAY** button.

Appendix C: Qualimetrics Transportable Automated Meteorological Station

This appendix provides useful information on the setup and use of TAMS units.

C.1 TAMS Unit Setup

1. Extend the tripod. Screw the tripod adapter on. Insert the met station onto the tripod adapter. Insert the anemometer and wind vane onto the appropriate sensors located at the top of the met station.
2. Use a compass to position the met station towards North as indicated on the figure shown on the side of the met station. Tighten the met station mounting clamp around the tripod adapter.
3. Connect the met station control cable to the bottom connector of the control/display unit.
4. Connect the RS232 output cable from the bottom connector of the control/display unit to the null modem adapter, to the HP 200 LX cable, then to the HP 200 LX.
5. Turn on the control/display unit.
6. Set the time on the control/display unit by pressing **SETUP** and the left arrow key until **Change Date/Time** is displayed and press **ENTER**. Type in the new time and press **ENTER**. Press the left arrow key again to set the date. Check the time and exit setup by pressing **SETUP**.
7. Check the data output settings by pressing **SETUP** and the left arrow key until **Change Data Output** is displayed and press **ENTER** four times. The display should now show **Computer Output is Enabled**. If it does not, press the left arrow key to toggle the setting to **Enabled**. Press **ENTER**.
8. The display should now show Output every 1 seconds. If it does not, press 1 to change the setting. Press **ENTER**. Press **SETUP** to exit setup.

Appendix D: Measurement Protocol and Logging Procedure

In order to facilitate orderly measurement of OSVs, the following protocol was instituted and found to be effective[*]:

- NPS and Volpe personnel determine what type of measurement will be conducted (constant speed or full throttle).
- NPS personnel specify vehicle to be measured.
- Volpe personnel note description of vehicle, check wind speed, and indicate readiness for measurement.
- NPS personnel indicate vehicle is beginning run.
- NPS personnel indicate vehicle speed to the OSV driver to allow the driver to maintain constant speed during passby[†].
- Volpe personnel[‡]:
    - note the rise and fall of the sound level on the LD824 as the vehicle travels through the measurement area,
    - note the maximum level displayed on the LD824,
    - confirm that the level at the start point and stop point were less than the maximum level by at least 10 dB, and
    - indicate any potentially contaminating sounds or wind levels greater than 12 mph, identifying the measurement as bad if appropriate.
- NPS personnel indicate if the vehicle strayed from a straight path along the travel path or if speed varied more than 2 mph from target speed, identifying the run as bad if appropriate.
- For the case of constant speed runs, NPS personnel indicate speed and Volpe personnel record the speed.
- Volpe personnel monitor the number of runs by each vehicle in each direction and announce when sufficient measurements are made for the specified vehicle under the specified operating condition.

---

[*] Communication between NPS and Volpe personnel was largely conducted by using radio transceivers. Radio silence was observed during measurement intervals.
[†] Vehicle speed was relayed to the OSV drivers by means of a lighted sign for vehicles traveling towards Grand Teton and by radio communication and hand signals for vehicles traveling towards Yellowstone.
[‡] For screening purposes ½ second L_{Aeq} were monitored rather than L_{ASmx}. During the post-processing and analysis, L_{ASmx} were used in accordance with the SAE standards.

Acoustics System Log

Volpe Center Acoustics Facility

Date:	Acoustics System:	Procedural Checklist: Adjust Calibration	Page ___ of ___
Site Name:	Aircraft:	*Record/Collect:*	
Site ID:	Personnel:	1) Cal Tone 4) Cal Tone 2 2) Mic Simulator 5) Data/Events 3) Pink Noise 6) End Cal Tone	

Event	Time (hh:mm:ss)		DAT ID	LD824 Gain/Range	LD824 Levels		DAT Input Level (dB)	dB Up/Down	L$_{max}$	Comments
	Start	End			SLM	RTA @ 1K	Ch 1			

Appendix E: Sound Level Time Histories

Sample time histories for each vehicle at idle, low, and high speed are provided in this appendix. One sample is provided for each test condition for each vehicle (provided the data are available). Low speed (15 mph) sound level time histories are provided in Figure 29 to Figure 38 for each snowcoach. High speed (30 mph) sound level time histories are provided in Figure 39 to Figure 43 for each snowcoach with available high speed data. Idle sound level time histories are provided in Figure 44 to Figure 53 for each snowcoach.

Low Speed Passby Time Histories

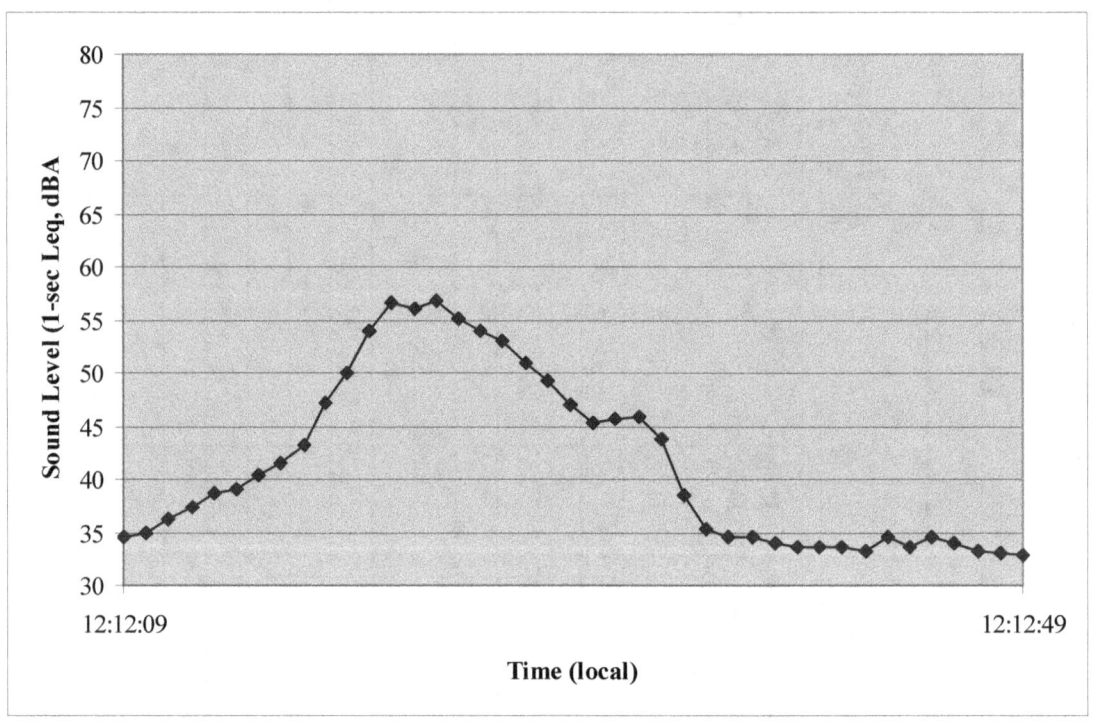

Figure 29. Alpen Guide, Right Side 15 mph (Feb 26th)

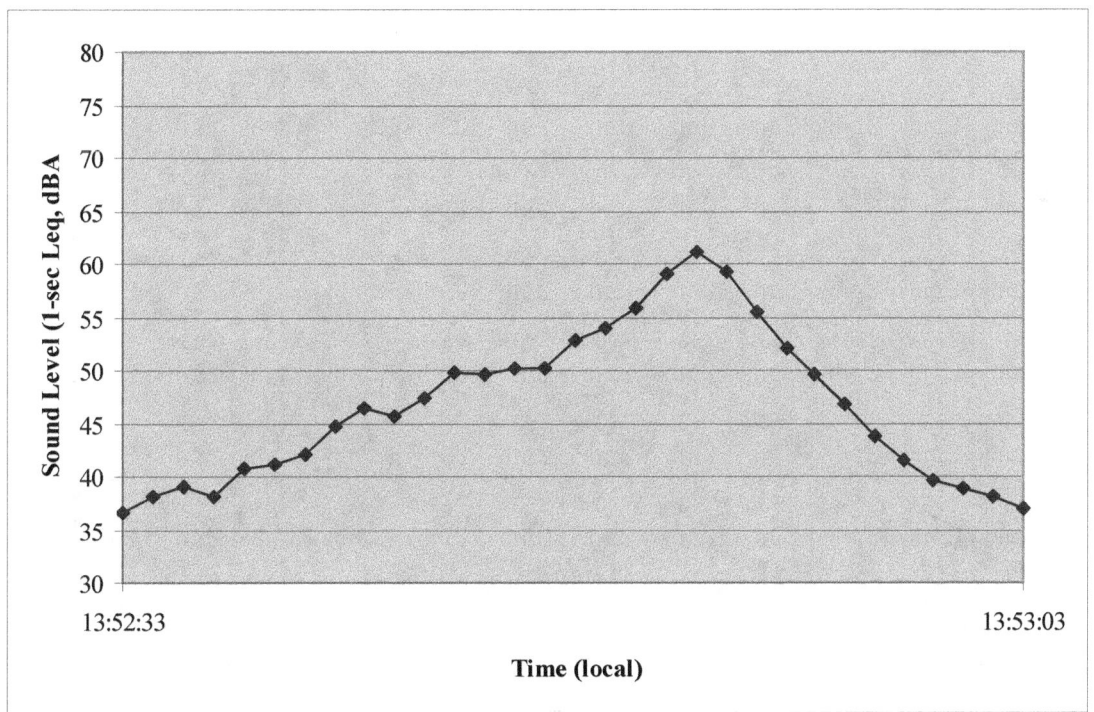

Figure 30. Goosewing Diesel Van, Right Side 15 mph (Feb 26th)

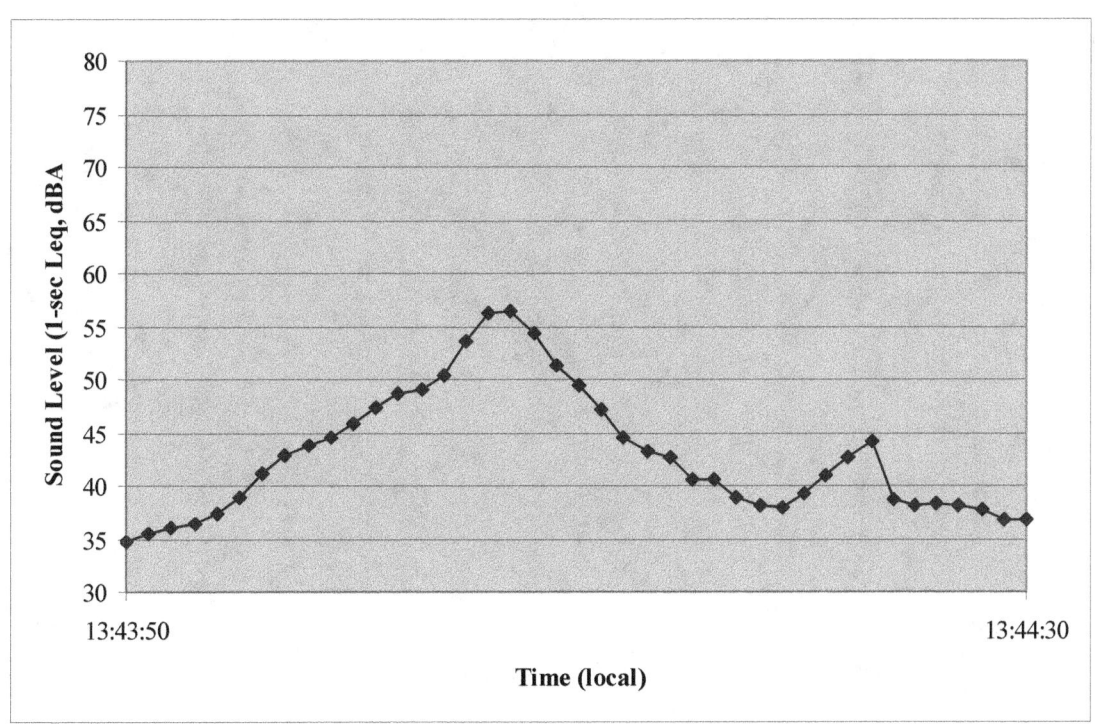

Figure 31. Rocky Mt. SC Tours, Right Side 15 mph (Feb 26th)

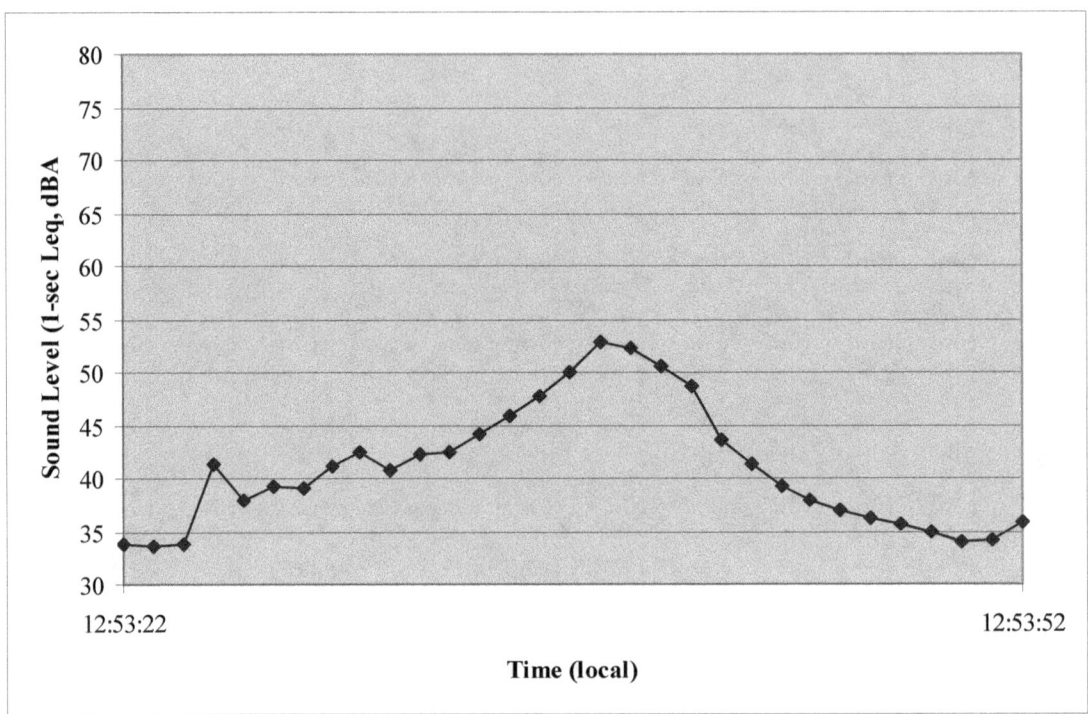

Figure 32. Yellowstone Expedition, Right Side 16 mph (Feb 26th)

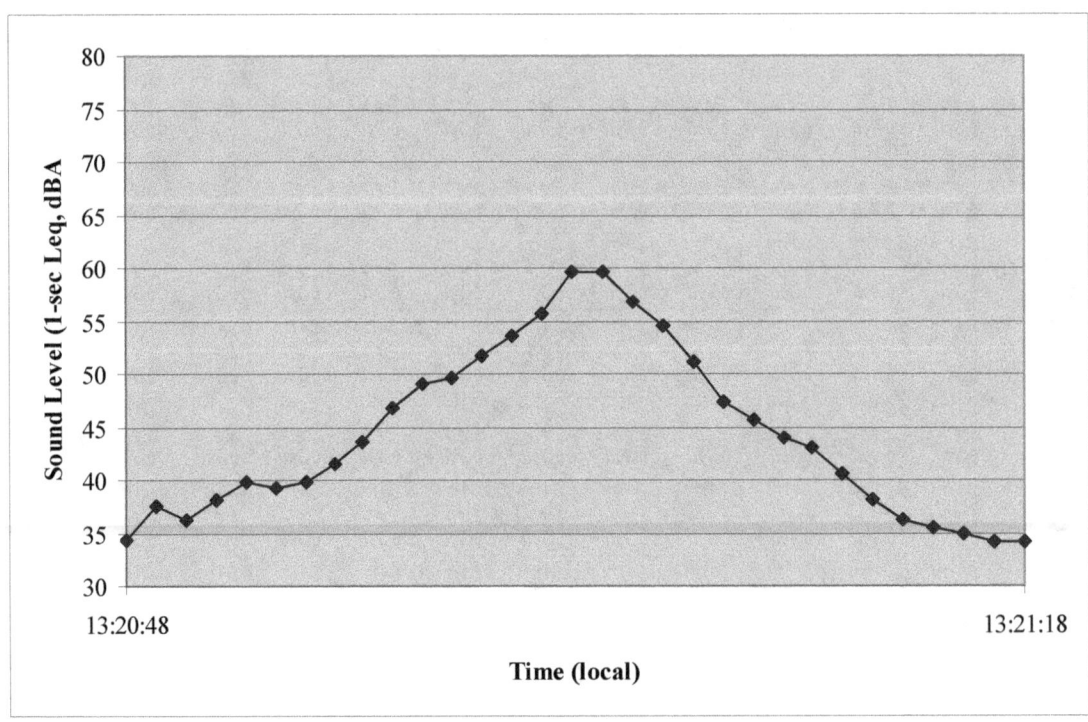

Figure 33. Yellowstone Snow Coach, Right Side 16 mph (Feb 26th)

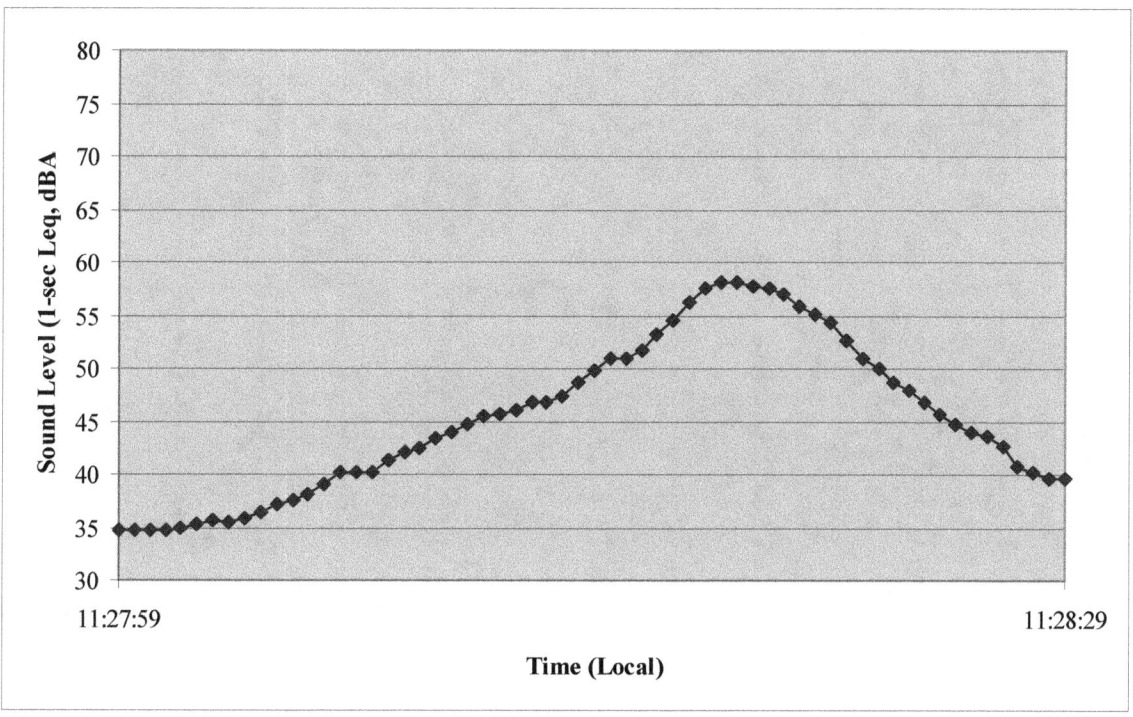

Figure 34. Xanterra 431, Right Side 15 mph (Feb 27th)

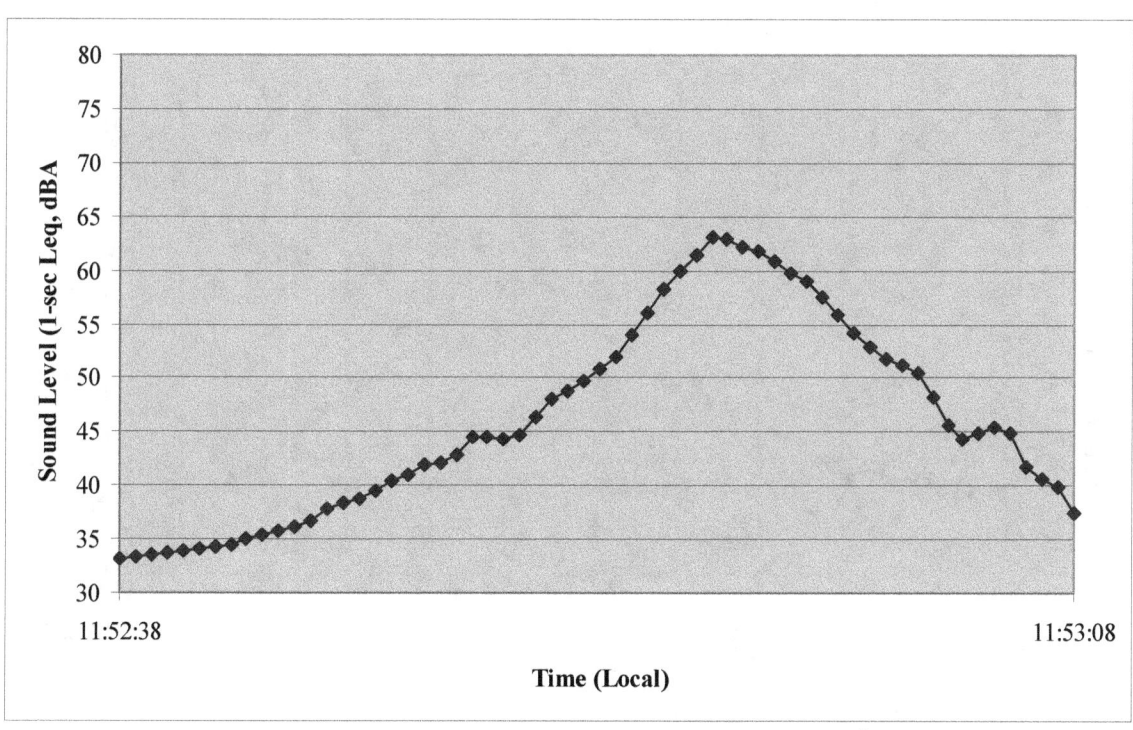

Figure 35. Xanterra 707, Right Side 15 mph (Feb 27th)

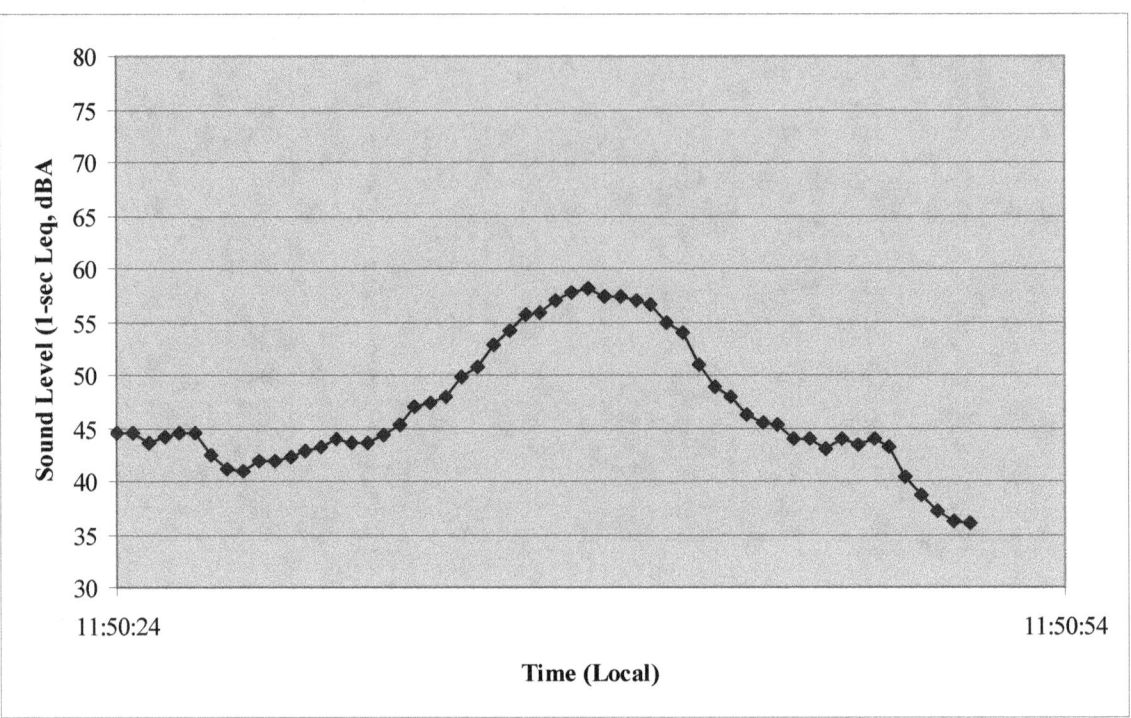

Figure 36. Xanterra 165, Left Side 15.5 mph (Feb 27th)

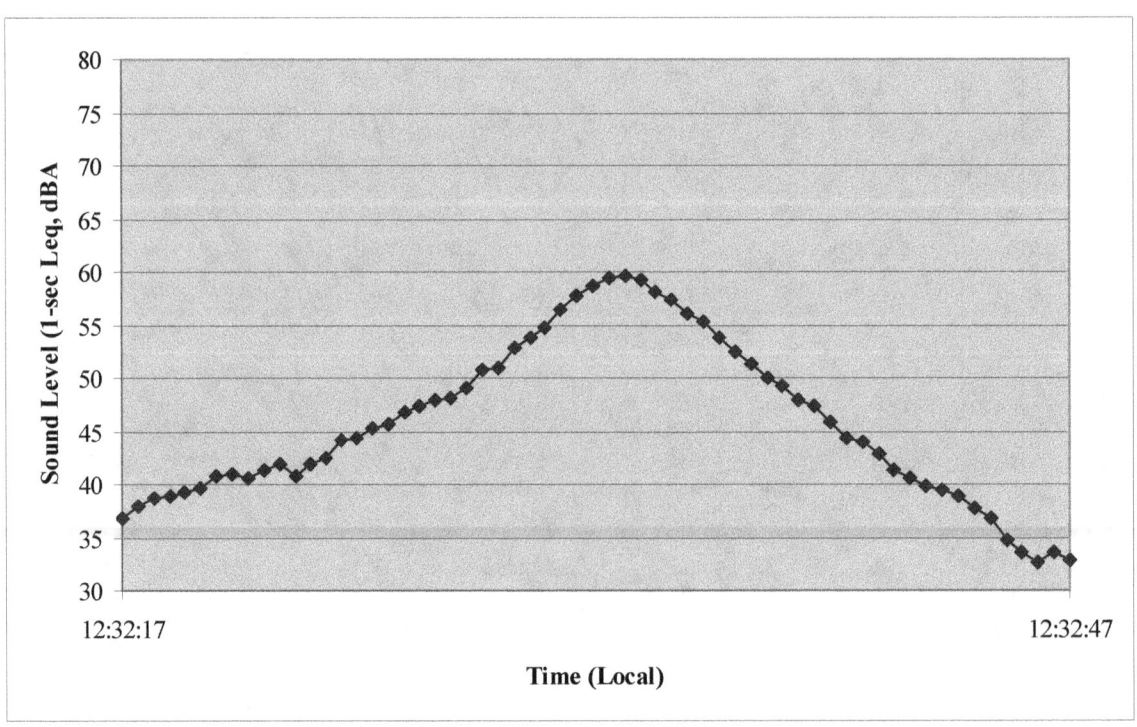

Figure 37. Goosewing Excursion, Right Side 16 mph (Feb 27th)

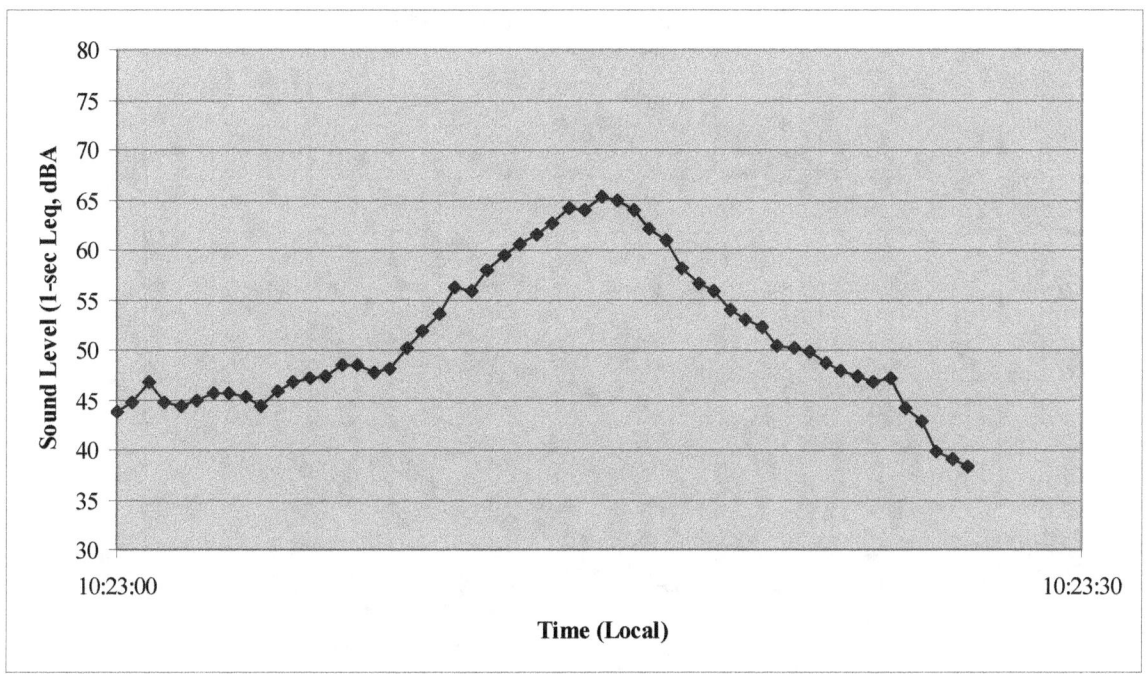

Figure 38. NPS Snow Coach, Left Side 15 mph (Feb 28th)

High Speed Passby Time Histories

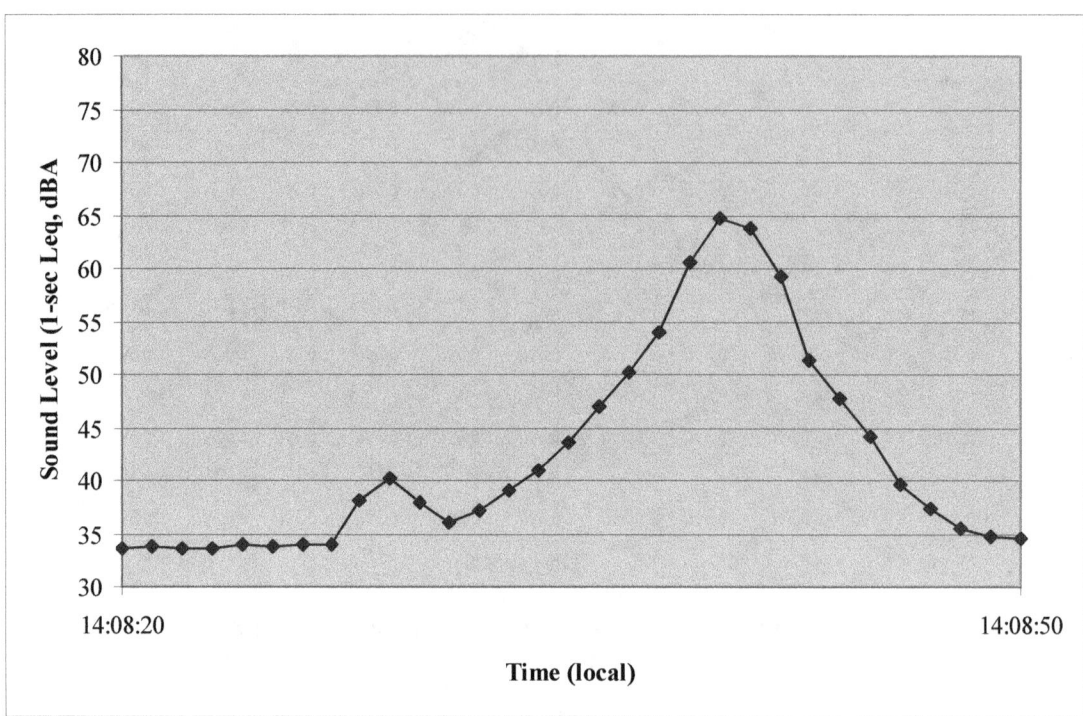

Figure 39. Alpen Guide, Right Side 34 mph (Feb 26th)

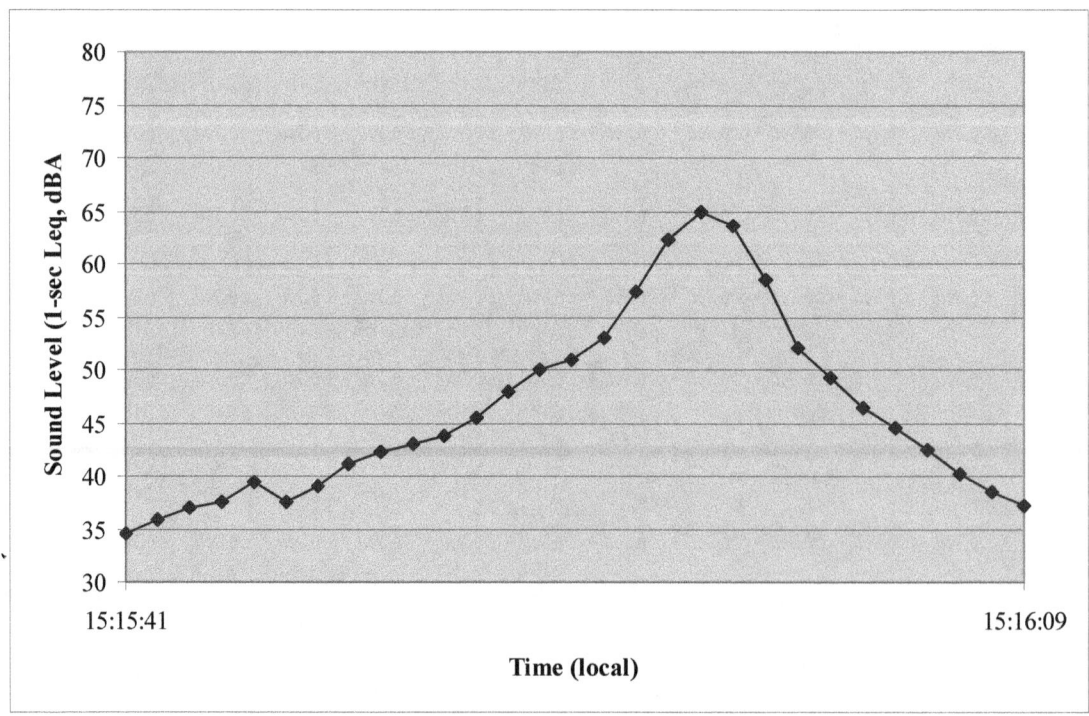

Figure 40. Goosewing Diesel Van, Left Side 28 mph (Feb 26th)

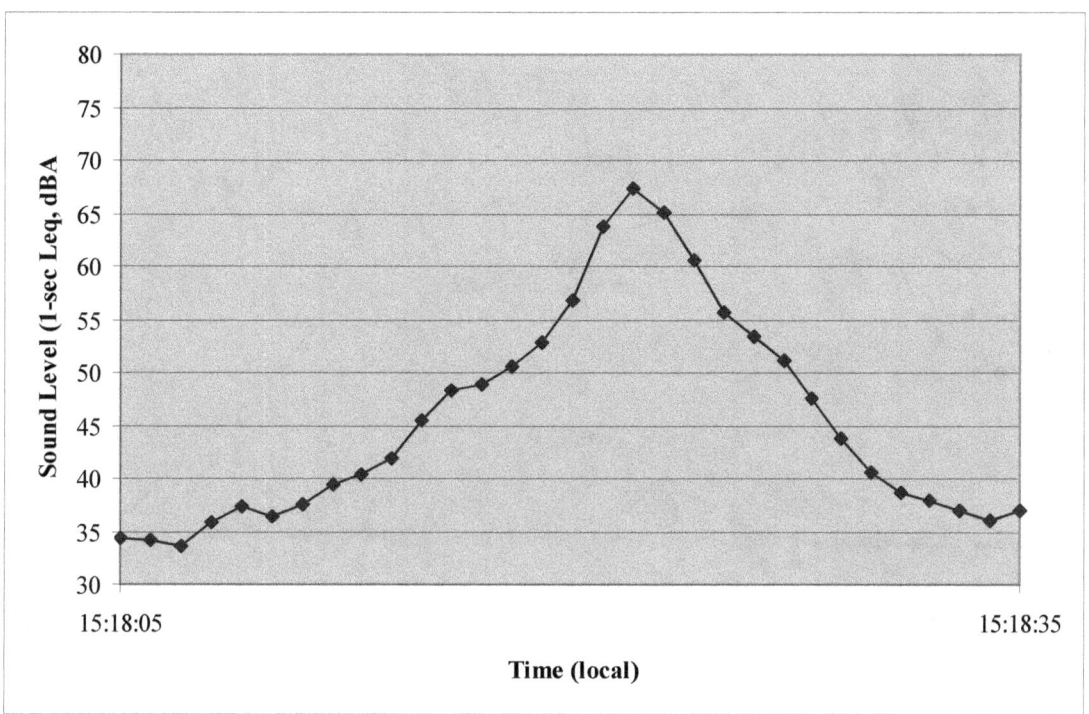

Figure 41. Rocky Mt. SC Tours, Left Side 29 mph (Feb 26th)

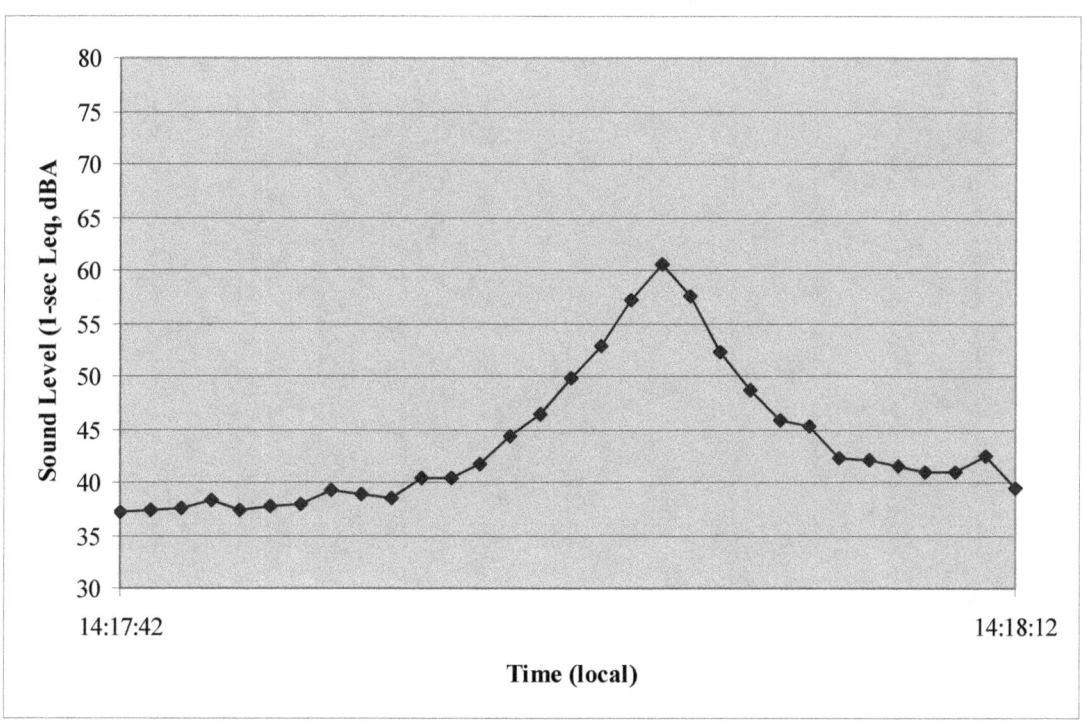

Figure 42. Yellowstone Expedition, Left Side 28.5 mph (Feb 26th)

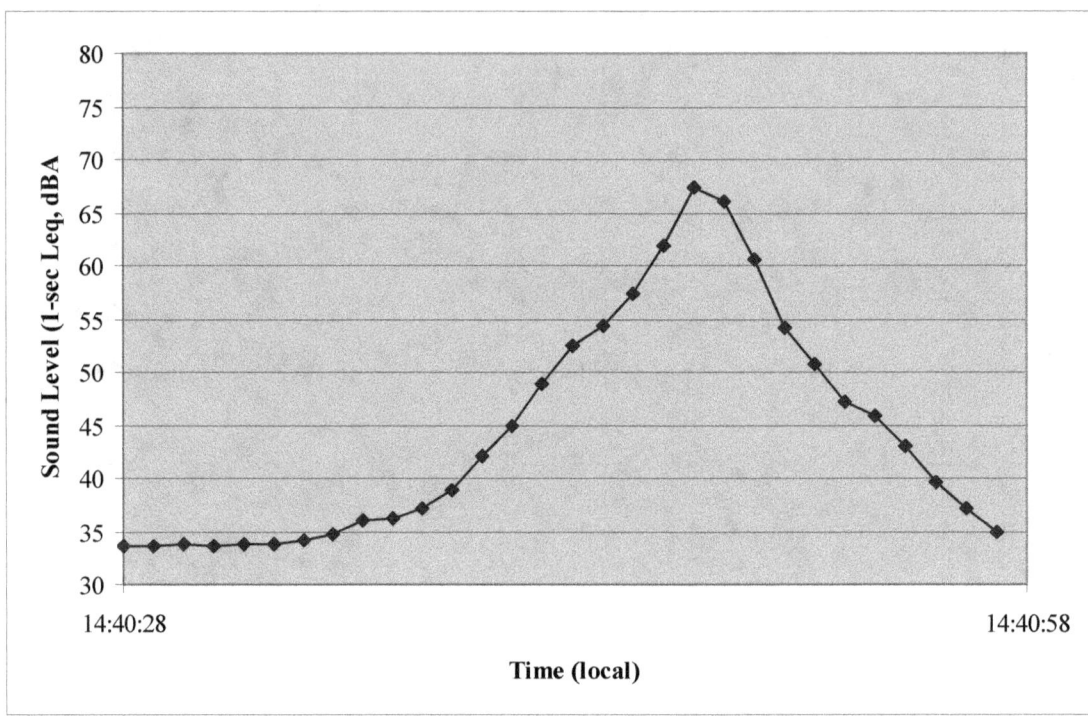

Figure 43. Yellowstone Snow Coach, Right Side 30 mph (Feb 26th)

Idle Time Histories

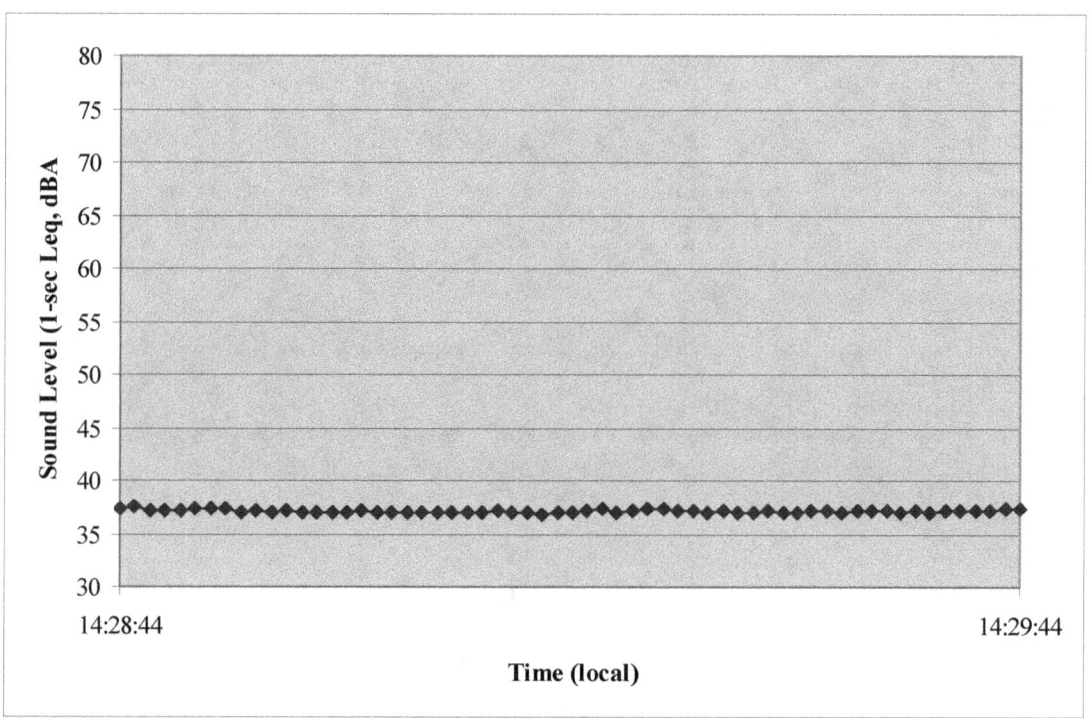

Figure 44. Alpen Guide, Right Side Idle (Feb 26th)

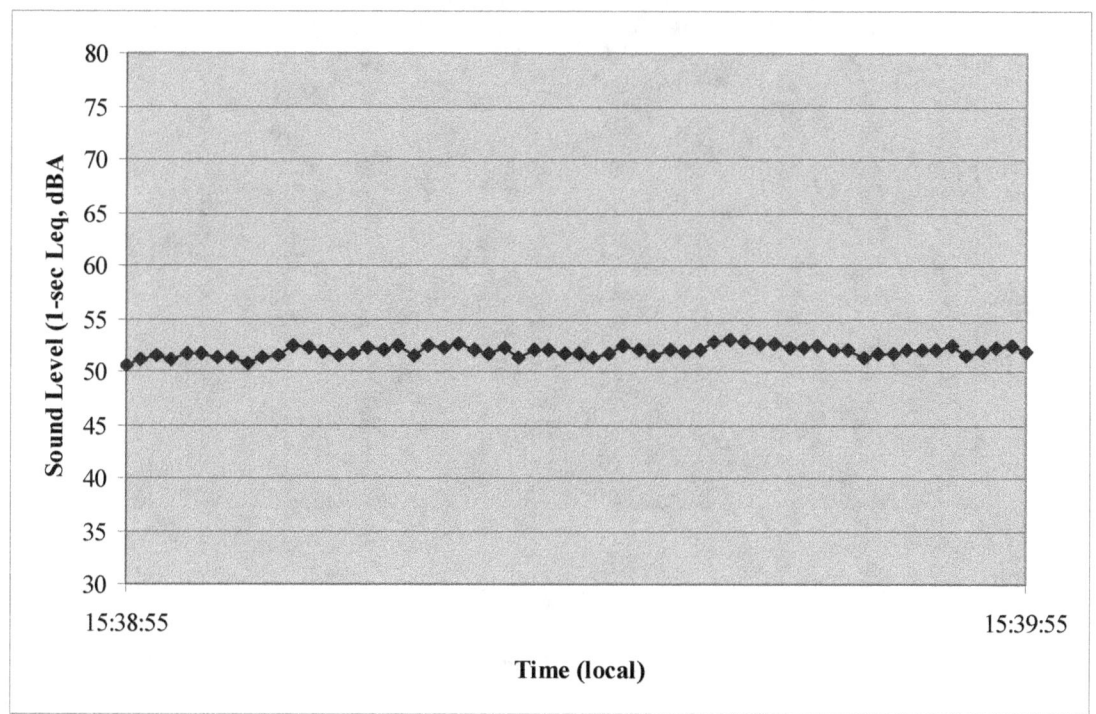

Figure 45. Goosewing Diesel Van, Left Side Idle (Feb 26th)

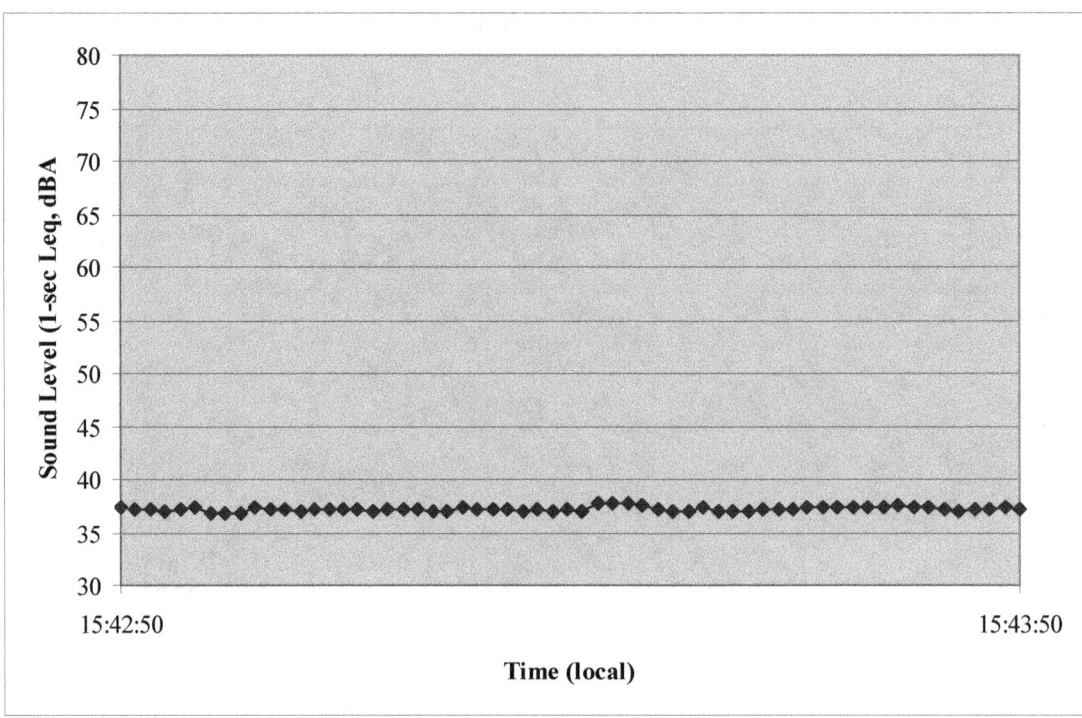

Figure 46. Rocky Mt. SC Tours, Left Side Idle (Feb 26th)

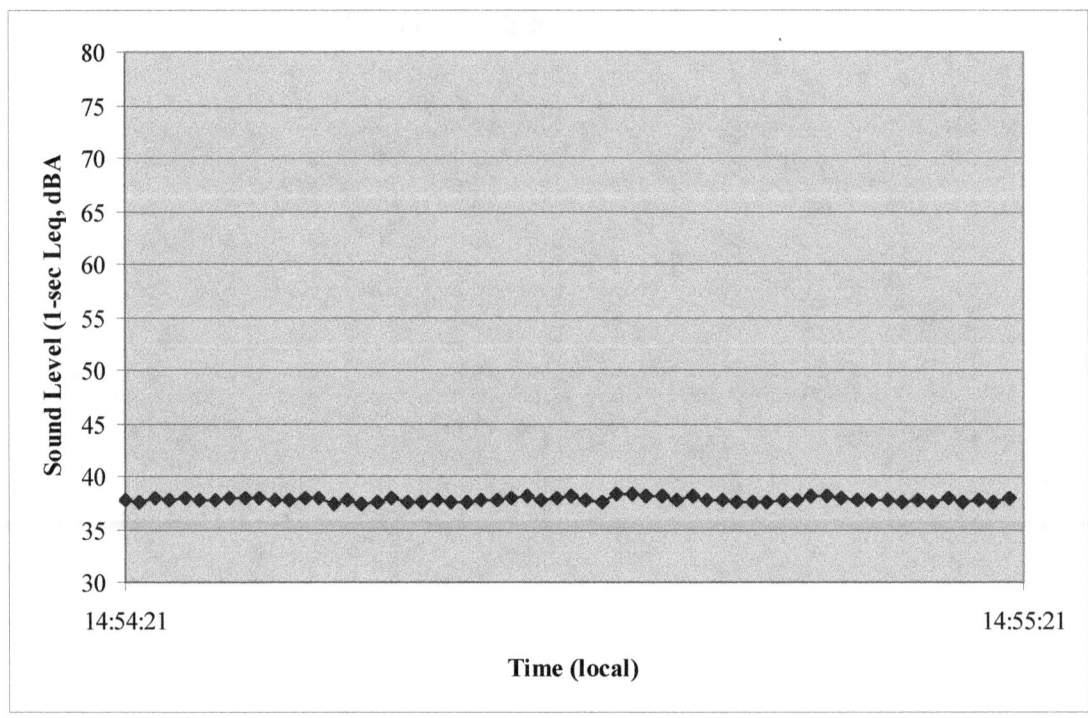

Figure 47. Yellowstone Expedition, Left Side Idle (Feb 26th)

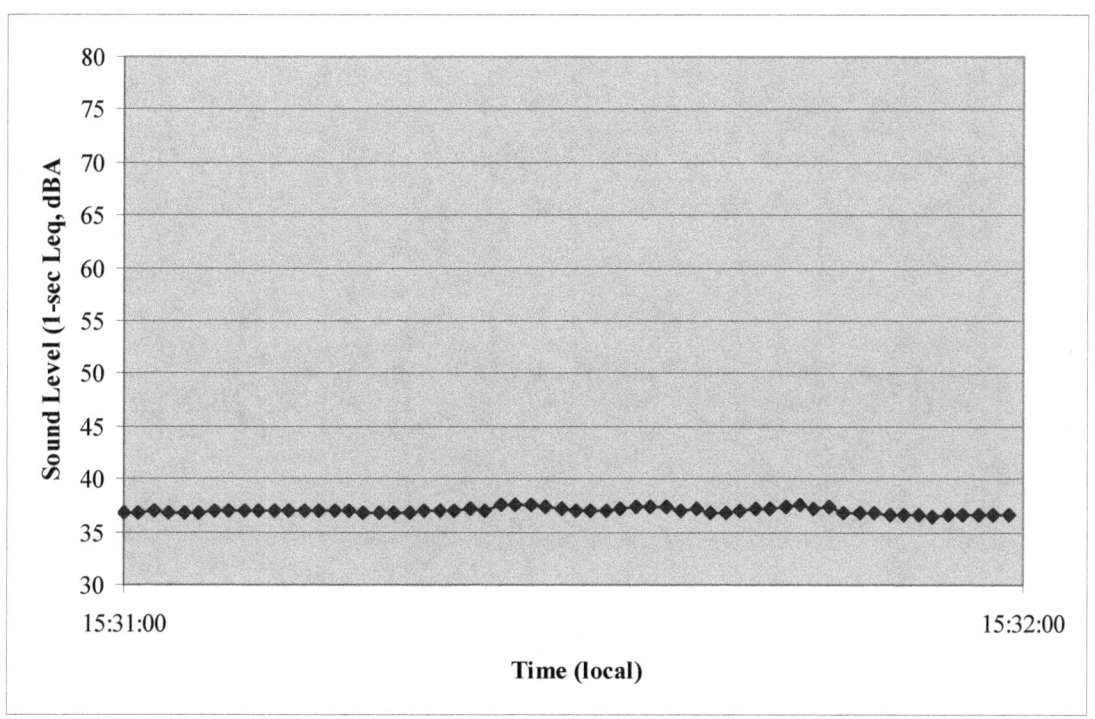

Figure 48. Yellowstone Snow Coach, Left Side Idle (Feb 26th)

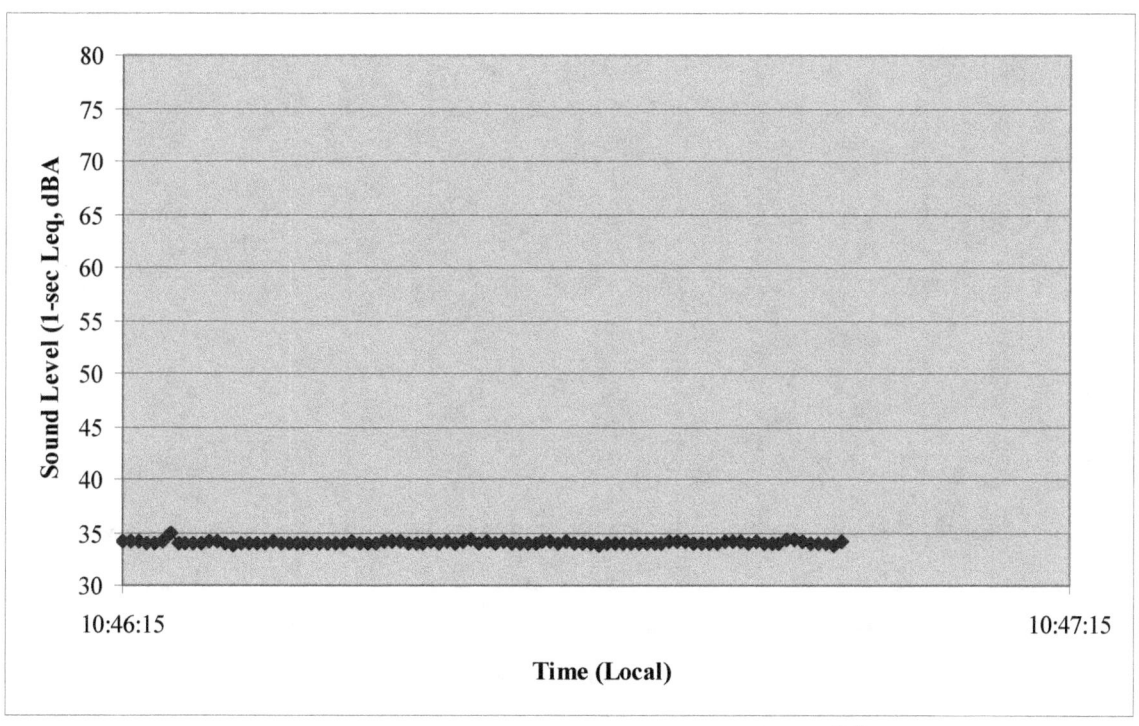

Figure 49. Xanterra 431, Right Side Idle (Feb 27th)*

---

* Measurement terminated early.

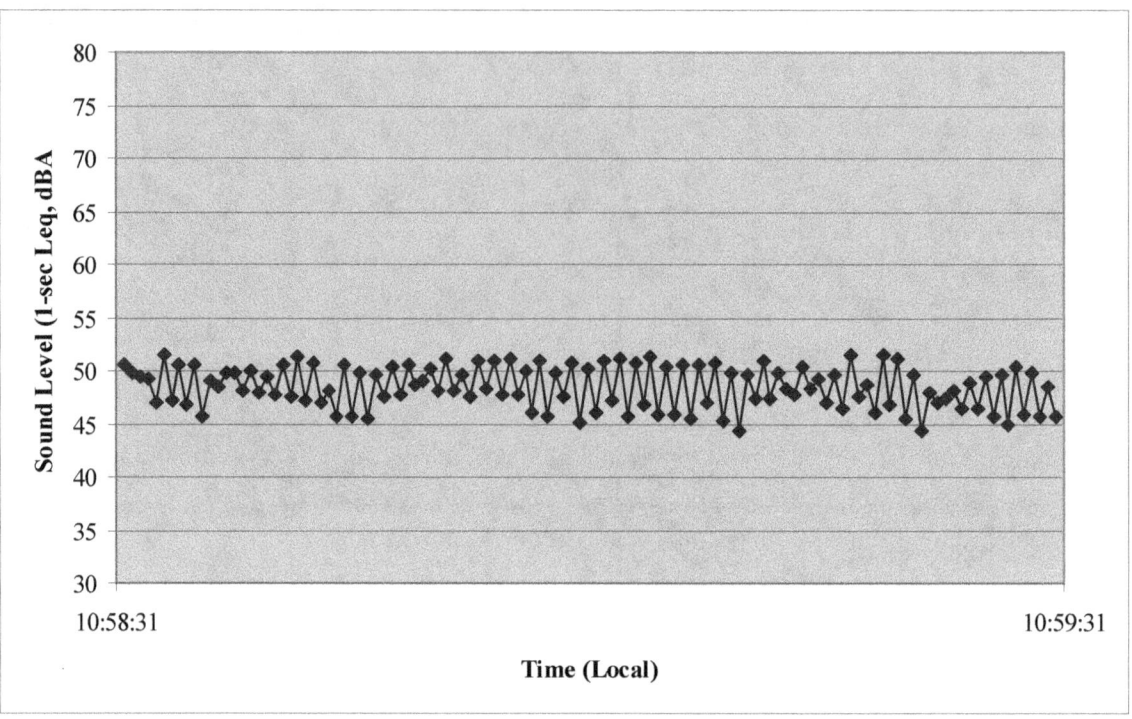

Figure 50. Xanterra 707, Right Side Idle (Feb 27th)

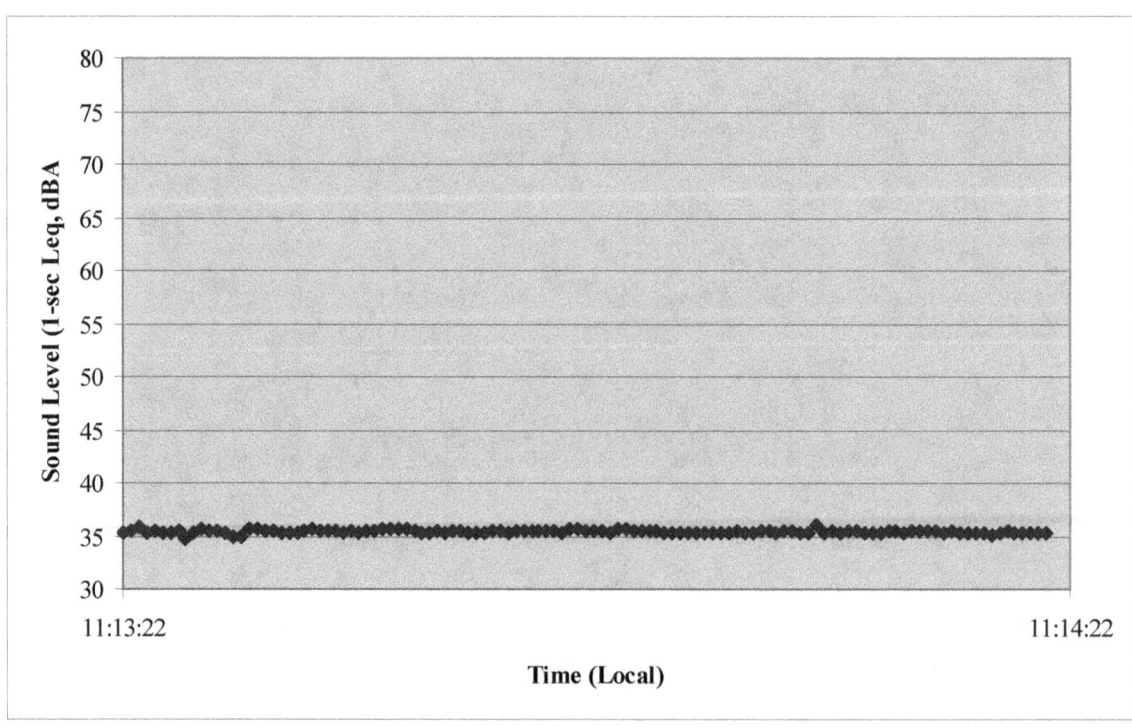

Figure 51. Xanterra 165, Left Side Idle (Feb 27th)

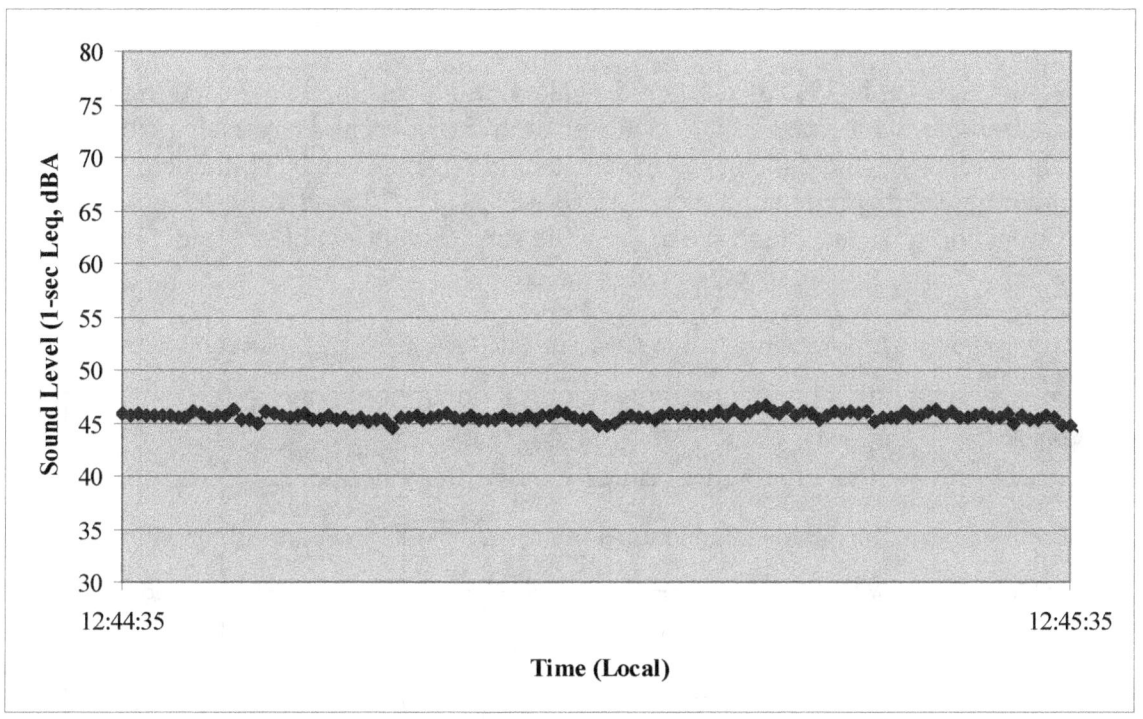

Figure 52. Goosewing Excursion, Left Side Idle (Feb 27th)

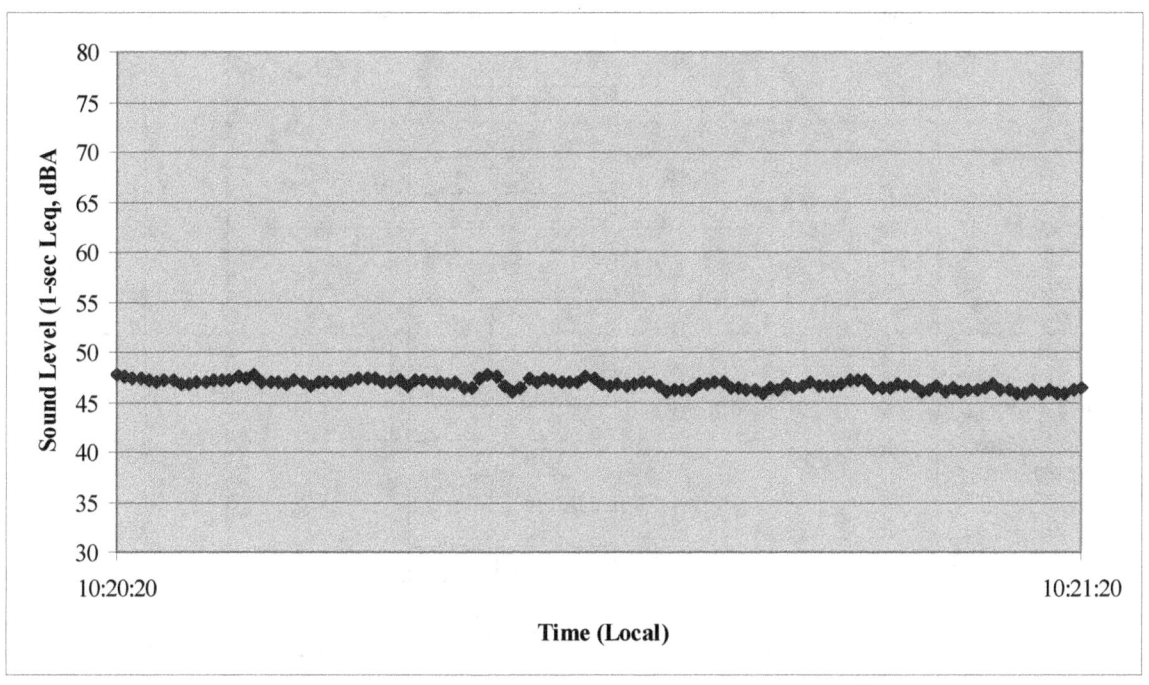

Figure 53. NPS Snow Coach, Left Side Idle (Feb 28th)

Appendix F: Overall Sound Levels

Table 21 through Table 25 provide the original overall levels that were used in the determination of the recorded L_{ASmx} / SEL for each snowcoach. Progressing from left to right data is averaged and culled until the final single overall value is obtained. The tables are organized by speed and metric. Table 21, Table 22, and Table 23 provide L_{ASmx} values for low speed, high speed, and idle respectively. Table 24 and Table 25 provide SEL values for low speed and high speed.

Table 21. Low Speed Measurements used to Generate Final Reported L_{ASmx} Level for each Snowcoach

Vehicle	Passby Speed, mph	Vehicle Side[#]	Passby Level (A-Max Slow), dB	Passby Average, dB	Overall Level, dB
Alpen Guide	Low Speed	Right	56.5	56.0	56.0
			57.0		
			54.6		
		Left	57.4	55.9	
			55.0		
			55.2		
Goosewing Diesel Van	Low Speed	Right	63.3	61.3	61.3
			60.1		
			60.5		
		Left	60.2	60.2	
			59.6		
			60.7		
Rocky Mt.	Low Speed	Right	56.9	56.3	56.3
			55.9		
			56.0		
		Left	56.2	55.8	
			56.6		
			54.7		
Yellowstone Expedition	Low Speed	Right	52.4	52.3	52.3
			52.1		
			52.3		
		Left	52.3	51.5	
			51.1		
			51.2		
			51.5		
Yellowstone SC Tours	Low Speed	Right	58.7	58.4	58.4
			57.6		
			59.4		
		Left	57.7	56.9	
			58.4		
			56.2		
			56.4		
			56.7		

[#] "Left" indicates left side of vehicle is closest to the microphones. "Right" indicates right side of vehicle is closest to the microphones.

Vehicle	Passby Speed, mph	Vehicle Side#	Passby Level (A-Max Slow), dB	Passby Average, dB	Overall Level, dB
Goosewing Excursion	Low Speed	Right	59.1	59.0	59.0
			58.8		
			59.1		
		Left	59.0	58.1	
			58.1		
			57.3		
			58.1		
Xanterra 165	Low Speed	Right	58.0	56.9	57.5
			56.4		
			56.8		
			56.2		
		Left	57.5	57.5	
			57.7		
			57.3		
			57.3		
Xanterra 431	Low Speed	Right	58.1	57.3	57.3
			56.0		
			57.7		
		Left	57.0	57.1	
			56.4		
			57.9		
Xanterra 707	Low Speed	Right	63.0	62.5	62.5
			62.2		
			62.2		
		Left	62.9	62.0	
			61.9		
			61.2		
NPS SC	Low Speed	Right	62.2	62.7	64.9
			62.6		
			63.2		
		Left	65.0	64.9	
			65.5		
			64.6		
			64.6		

"Left" indicates left side of vehicle is closest to the microphones. "Right" indicates right side of vehicle is closest to the microphones.

Table 22. High Speed Measurements used to Generate Final Reported L_{ASmx} Level for each Snowcoach

Vehicle	Passby Speed, mph	Vehicle Side[#]	Passby Level (A-Max Slow), dB	Passby Average, dB	Overall Level, dB
Alpen Guide	High Speed	Right	64.1	63.6	63.6
			64.2		
			62.6		
		Left	63.0	63.1	
			63.9		
			62.5		
Goosewing Diesel Van	High Speed	Right	64.0	64.1	64.7
			63.5		
			64.8		
		Left	64.2	64.7	
			65.2		
			64.7		
Rocky Mt.	High Speed	Right	65.4	67.5	67.9
			68.1		
			68.9		
		Left	66.2	67.9	
			68.9		
			68.7		
Yellowstone Expedition	High Speed	Right	58.2	57.2	59.1
			56.5		
			57.0		
		Left	57.9	59.1	
			59.6		
			59.9		
Yellowstone SC Tours	High Speed	Right	66.0	65.8	65.8
			64.8		
			66.6		
		Left	65.8	65.4	
			64.9		
			65.5		
			65.2		
Goosewing Excursion	High Speed	N/A			
Xanterra 165	High Speed	N/A			
Xanterra 431	High Speed	N/A			
Xanterra 707	High Speed	N/A			
NPS SC	High Speed	N/A			

[#] "Left" indicates left side of vehicle is closest to the microphones. "Right" indicates right side of vehicle is closest to the microphones.

Table 23. Idle Measurements used to Generate Final Reported L_{Aeq} Level for each Snowcoach

Vehicle	Passby Speed, mph	Vehicle Side[#]	Pass-by Average L_{Aeq}, dB	Overall Level, dB
Alpen Guide	Idle	Right	38.6	38.6
		Left	36.7	
Goosewing Diesel Van	Idle	Right	52.2	53.0
		Left	53.0	
Rocky Mt.	Idle	Right	37.7	37.7
		Left	37.7	
Yellowstone Expedition	Idle	Right	36.9	38.8
		Left	38.8	
Yellowstone SC Tours	Idle	Right	37.2	37.5
		Left	37.5	
Goosewing Excursion	Idle	Right	46.4	47.2
		Left	47.2	
Xanterra 165	Idle	Right	35.0	36.4
		Left	36.4	
Xanterra 431	Idle	Right	34.4	34.4
		Left	34.3	
Xanterra 707	Idle	Right	50.2	50.2
		Left	42.4	
NPS SC	Idle	Right	46.4	47.7
		Left	47.7	

[#] "Left" indicates left side of vehicle is closest to the microphones. "Right" indicates right side of vehicle is closest to the microphones.

Table 24. Low Speed Measurements used to Generate Final Reported SELs for each Snowcoach

Vehicle	Passby Speed, mph	Vehicle Side[#]	Passby Level (SEL), dB	Passby Average, dB	Overall Level, dB
Alpen Guide	Low Speed	Right	64.5	63.8	63.8
			64.5		
			62.0		
		Left	64.1	62.7	
			61.8		
			61.8		
Goosewing Diesel Van	Low Speed	Right	68.6	67.2	67.2
			66.1		
			66.5		
		Left	66.6	66.7	
			66.1		
			67.2		
Rocky Mt.	Low Speed	Right	63.7	63.3	63.3
			63.2		
			63.0		
		Left	63.2	63.1	
			63.7		
			62.1		
Yellowstone Expedition	Low Speed	Right	59.2	59.1	59.1
			58.8		
			59.4		
		Left	59.3	58.7	
			58.1		
			58.3		
			59.1		
Yellowstone SC Tours	Low Speed	Right	65.6	65.2	65.2
			64.6		
			65.6		
			64.9		
		Left	65.6	64.2	
			63.9		
			63.6		
			63.4		

[#] "Left" indicates left side of vehicle is closest to the microphones. "Right" indicates right side of vehicle is closest to the microphones.

Vehicle	Passby Speed, mph	Vehicle Side[#]	Passby Level (SEL), dB	Passby Average, dB	Overall Level, dB
Goosewing Excursion	Low Speed	Right	65.7	65.9	65.9
			65.8		
			66.1		
		Left	66.2	65.3	
			65.1		
			64.4		
			65.2		
Xanterra 165	Low Speed	Right	65.0	64.2	64.7
			64.0		
			64.1		
		Left	63.5	64.7	
			64.8		
			65.1		
			64.6		
			64.4		
Xanterra 431	Low Speed	Right	65.4	64.7	65.0
			63.0		
			65.2		
		Left	65.2	65.0	
			64.6		
			65.1		
Xanterra 707	Low Speed	Right	69.5	69.1	69.1
			68.8		
			68.9		
		Left	69.7	68.9	
			68.6		
			68.1		
NPS SC	Low Speed	Right	64.9	65.3	66.5
			64.9		
			65.9		
		Left	66.5	66.5	
			66.3		
			66.1		
			67.2		

[#] "Left" indicates left side of vehicle is closest to the microphones. "Right" indicates right side of vehicle is closest to the microphones.

Table 25. **High Speed Measurements used to Generate Final Reported SELs for each Snowcoach**

Vehicle	Passby Speed, mph	Vehicle Side[#]	Passby Level (SEL), dB	Passby Average, dB	Overall Level, dB
Alpen Guide	High Speed	Right	68.6	68.4	68.4
			68.8		
			67.6		
		Left	67.8	68.0	
			68.7		
			67.3		
Goosewing Diesel Van	High Speed	Right	69.0	69.2	70.0
			68.9		
			69.7		
		Left	69.4	70.0	
			70.8		
			69.8		
Rocky Mt.	High Speed	Right	70.8	72.3	72.7
			72.5		
			73.2		
		Left	71.1	72.7	
			73.7		
			73.0		
Yellowstone Expedition	High Speed	Right	64.1	63.1	64.0
			62.1		
			62.7		
		Left	63.3	64.0	
			64.4		
			64.3		
Yellowstone SC Tours	High Speed	Right	70.6	70.5	70.5
			69.6		
			71.1		
		Left	70.6	70.3	
			69.9		
			70.3		
			70.2		
Goosewing Excursion	High Speed	N/A			
Xanterra 165	High Speed	N/A			
Xanterra 431	High Speed	N/A			
Xanterra 707	High Speed	N/A\			
NPS SC	High Speed	N/A			

[#] "Left" indicates left side of vehicle is closest to the microphones. "Right" indicates right side of vehicle is closest to the microphones.

Appendix G: One-Third Octave Band Levels

One-third octave band levels associated with the low speed passby time histories found in Appendix E are provided numerically in Table 26 and graphically in Figure 54 to Figure 63. One-third octave band levels associated with high speed passby time histories found in Appendix E are provided numerically in Table 27 and graphically in Figure 64 to Figure 68.

One-Third Octave Band Results at Low Speeds (Nominally 15 mph)

Table 26. Maximum One-Third Octave Band Levels for Select Events at Low Speed (Nominally 15 mph), dB

Vehicle	Vehicle Side#	Start Time	50	63	80	100	125	160	200	250	315	400	500	630	800	1000	1250	1600	2000	2500	3150	4000	5000	6300	8000	10000
NPS SC	Left	10:18:26	60.6	54.4	57.2	61.1	51.8	58.7	54.1	51.8	52.9	54.2	54.2	54.5	54.4	54.9	56.1	54.4	52.7	52.6	51.5	48.3	46.4	44.6	43.6	38.3
Xanterra 707	Right	11:52:32	60.3	60.3	63.7	65.0	59.0	65.6	62.3	54.6	54.1	51.9	52.4	52.8	53.1	51.2	48.9	48.2	49.0	47.4	45.5	45.4	45.4	41.1	39.5	37.5
Goose wing	Right	13:52:00	57.6	52.4	63.0	49.6	53.5	55.4	48.2	52.7	51.6	48.2	46.2	46.4	46.4	50.5	52.4	49.5	46.7	47.7	46.6	42.4	43.3	41.8	41.4	37.9
YS SC Tour	Right	13:21:00	50.1	50.2	64.3	49.1	45.9	56.0	50.2	55.8	57.5	53.5	53.3	51.9	51.2	49.8	47.6	43.7	41.7	39.9	38.2	37.6	36.0	36.0	36.1	35.3
Goose wing	Right	12:32:15	46.8	51.5	64.2	51.5	45.6	52.6	54.4	53.3	52.6	49.5	49.0	49.0	48.4	49.5	49.2	48.6	46.9	46.7	45.9	42.2	41.7	39.8	37.7	34.7
Xanterra 431	Right	11:27:53	62.9	54.4	62.2	52.4	56.2	53.9	59.8	53.0	52.8	51.4	48.5	47.5	47.8	46.4	45.6	43.7	44.4	42.7	42.1	38.8	36.9	34.6	33.0	31.4
Xanterra 165	Left	11:50:24	58.5	61.0	53.4	54.6	50.7	55.8	51.2	46.5	49.3	46.8	48.3	50.4	49.3	49.3	46.7	45.4	44.2	42.2	40.9	40.6	39.7	38.4	36.5	33.8
Alpen Guide	Right	12:12:29	51.0	61.0	56.4	60.0	51.3	49.3	48.5	46.9	46.6	45.2	44.8	45.9	47.7	49.1	49.5	45.0	43.7	43.3	38.8	36.3	35.5	31.0	28.1	25.6
Rocky Mt	Right	13:43:00	50.2	52.2	60.9	49.7	46.2	57.9	49.6	52.7	53.3	51.0	50.0	48.1	46.8	44.0	41.4	39.6	38.5	41.8	37.3	33.2	34.1	31.9	30.4	26.2
YS Exp.	Right	12:53:00	53.5	54.9	62.2	52.1	54.7	51.3	50.0	48.3	46.1	44.6	43.8	43.6	42.0	41.2	41.7	39.5	37.3	35.3	33.5	32.4	30.8	27.8	26.6	24.0

"Left" indicates left side of vehicle is closest to the microphones. "Right" indicates right side of vehicle is closest to the microphones.

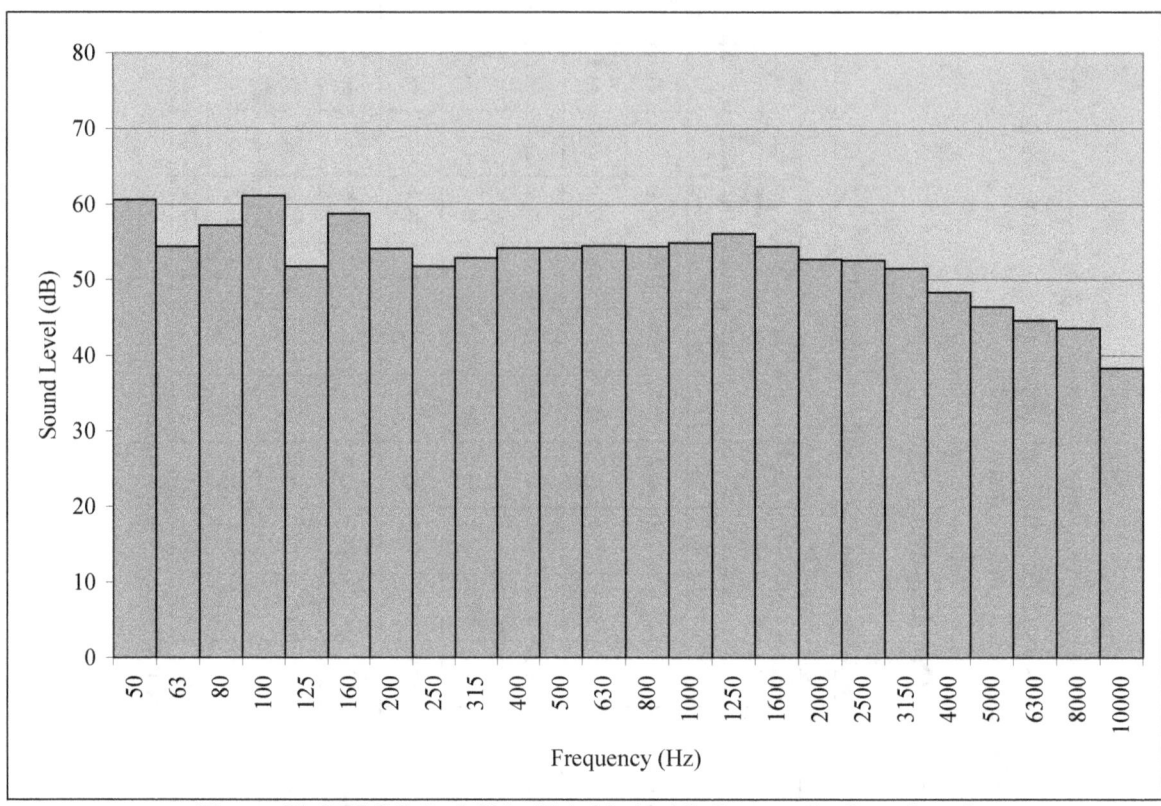

Figure 54. NPS SC, Left Side (Feb 28, 10:18) Spectra for Low Speed

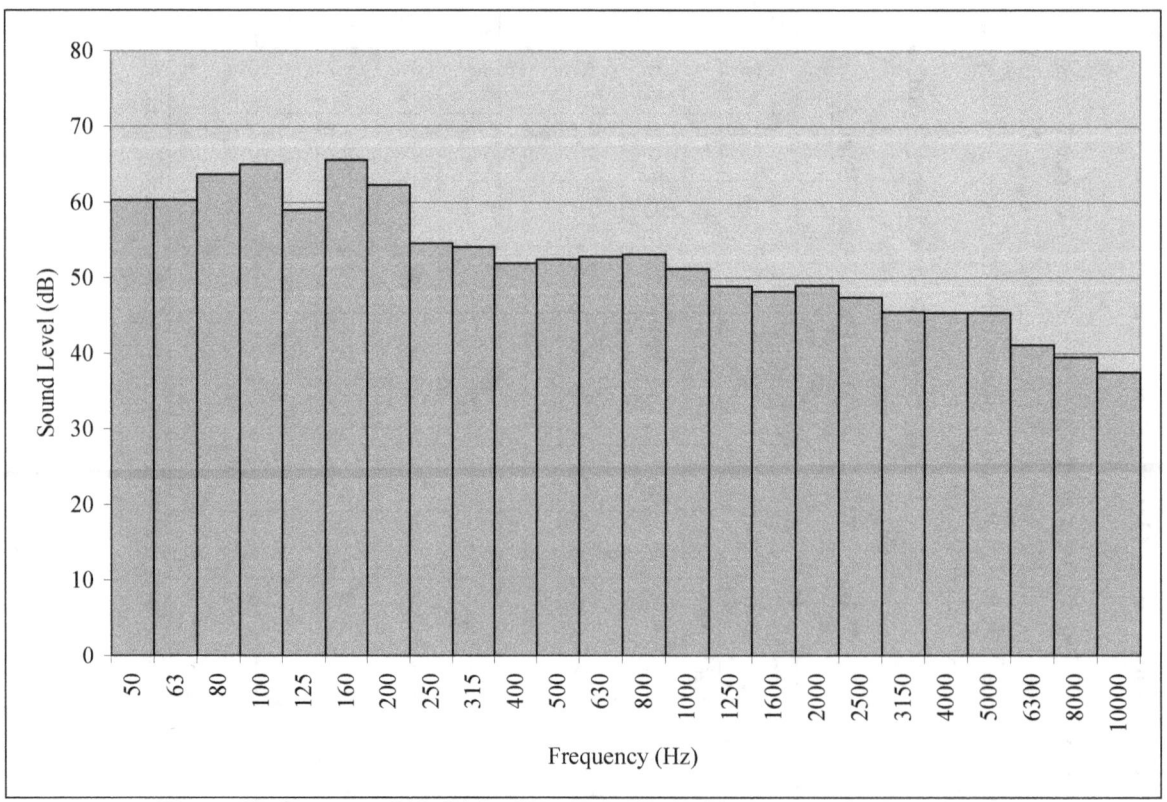

Figure 55. Xanterra 707, Right Side (Feb 27, 11:52) Spectra for Low Speed

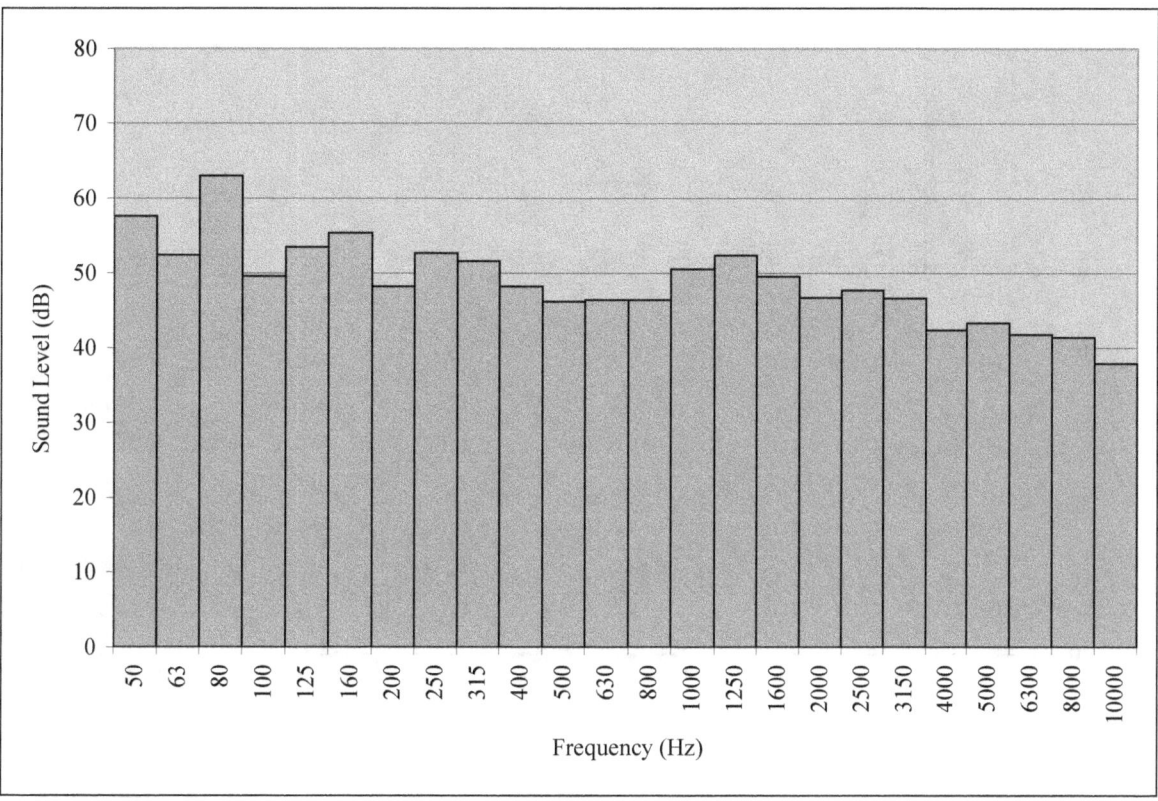

Figure 56. Goosewing Diesel Van, Right Side (Feb 26, 13:52) Spectra for Low Speed

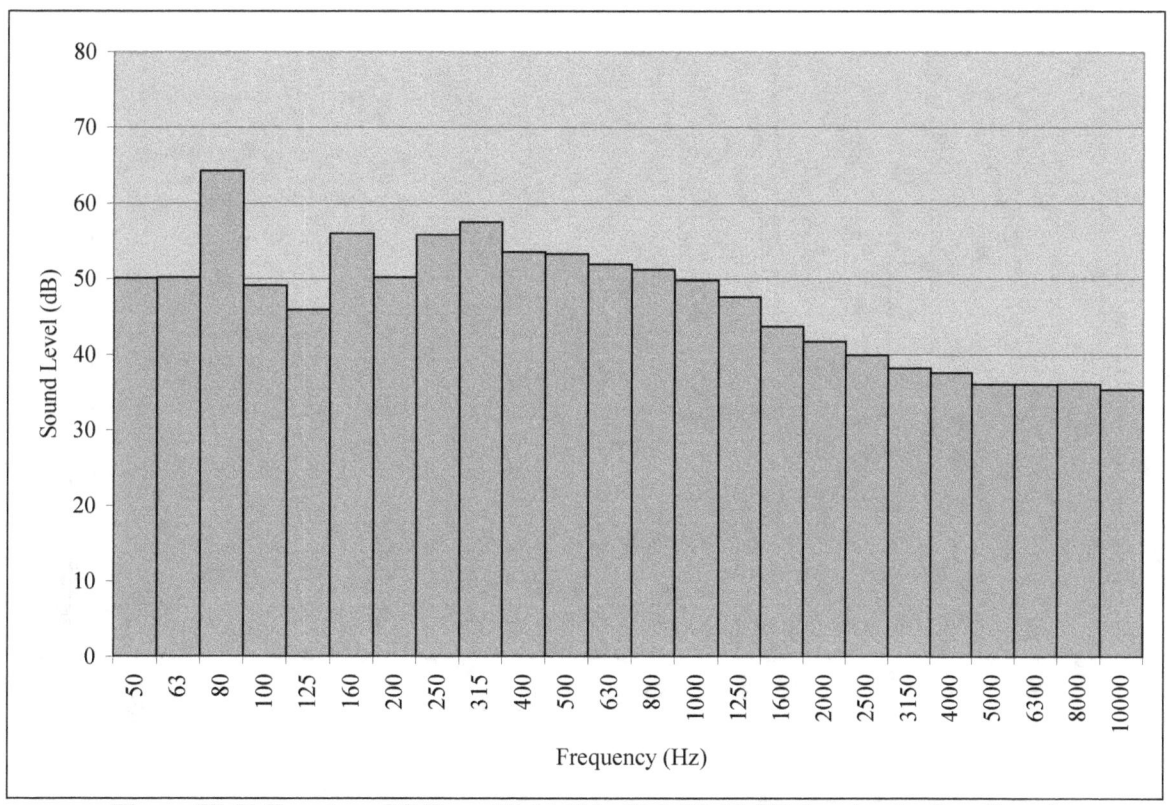

Figure 57. Yellowstone SC Tour, Right Side (Feb 26, 13:21) Spectra for Low Speed

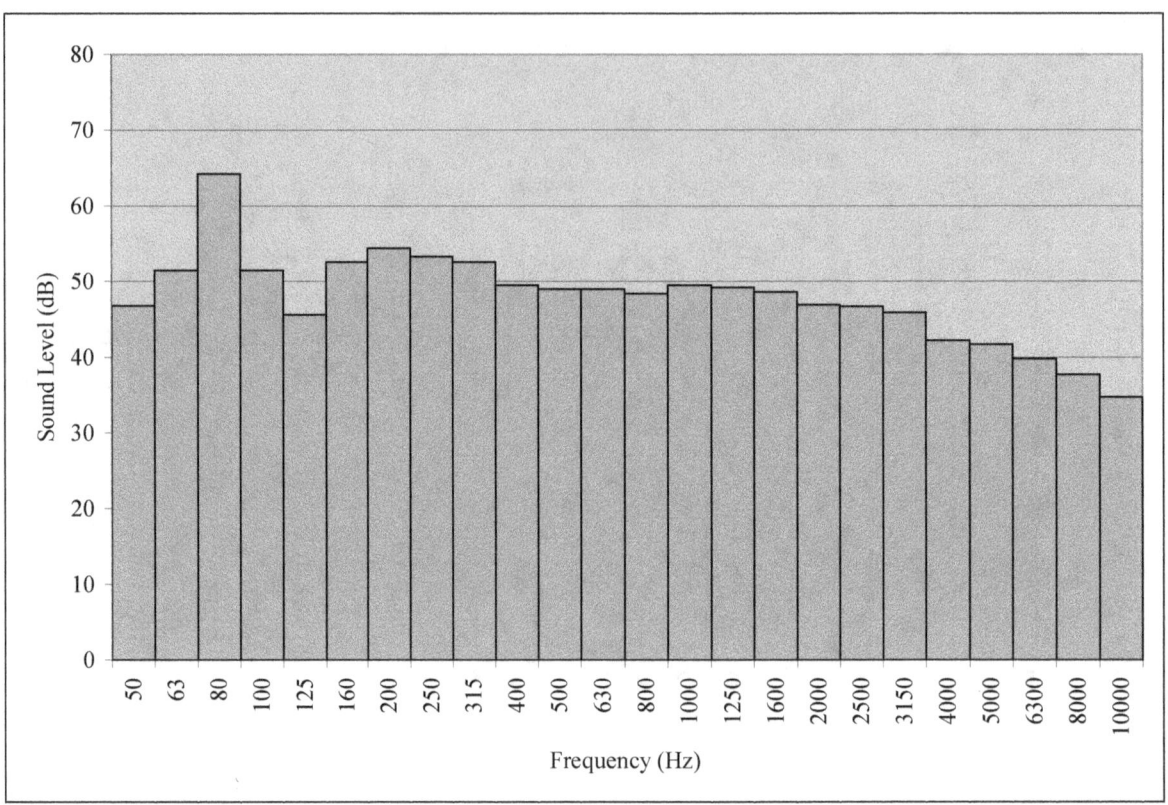

Figure 58. Goosewing Excursion, Right Side (Feb 27, 12:32) Spectra for Low Speed

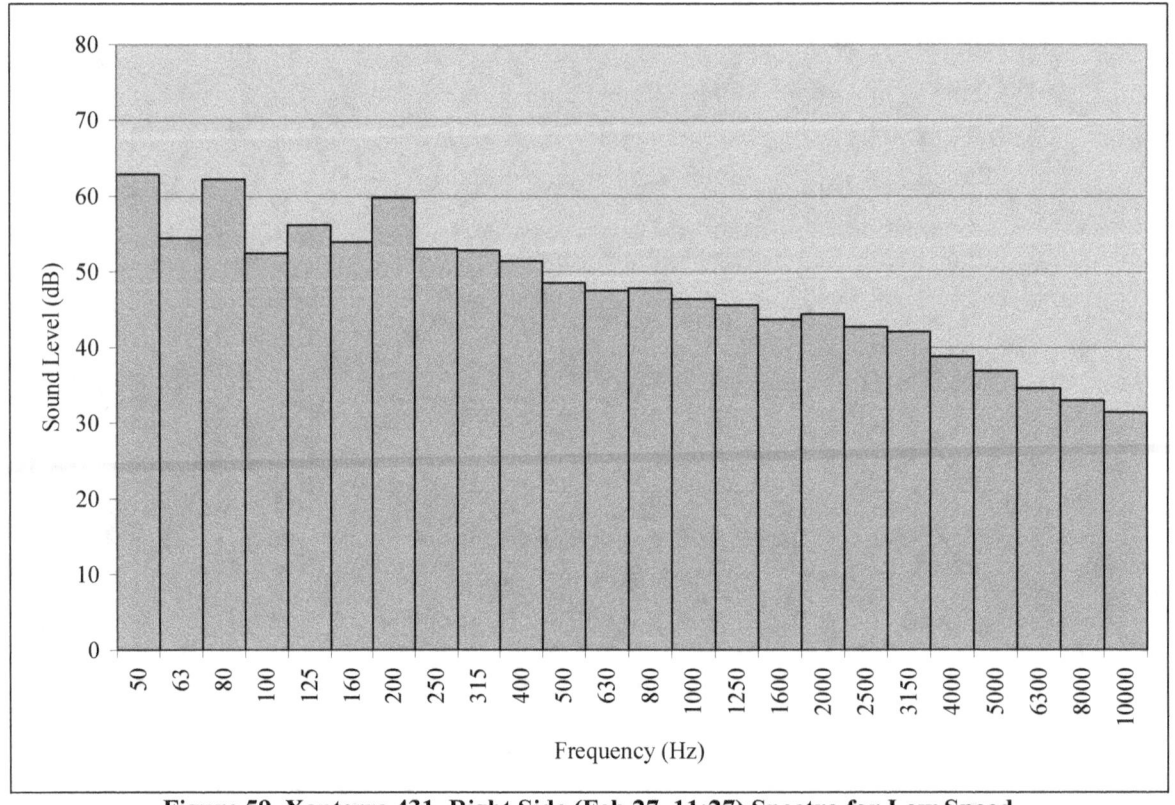

Figure 59. Xanterra 431, Right Side (Feb 27, 11:27) Spectra for Low Speed

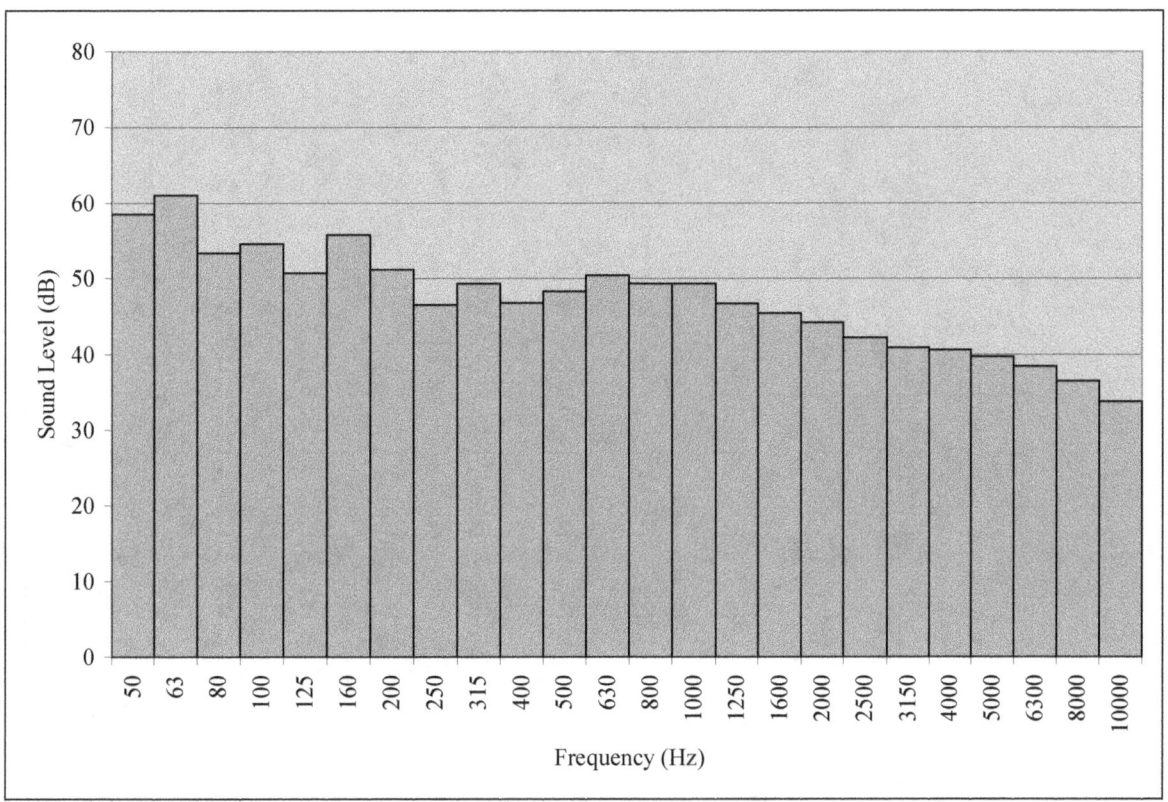

Figure 60. Xanterra 165, Left Side (Feb 27, 11:50) Spectra for Low Speed

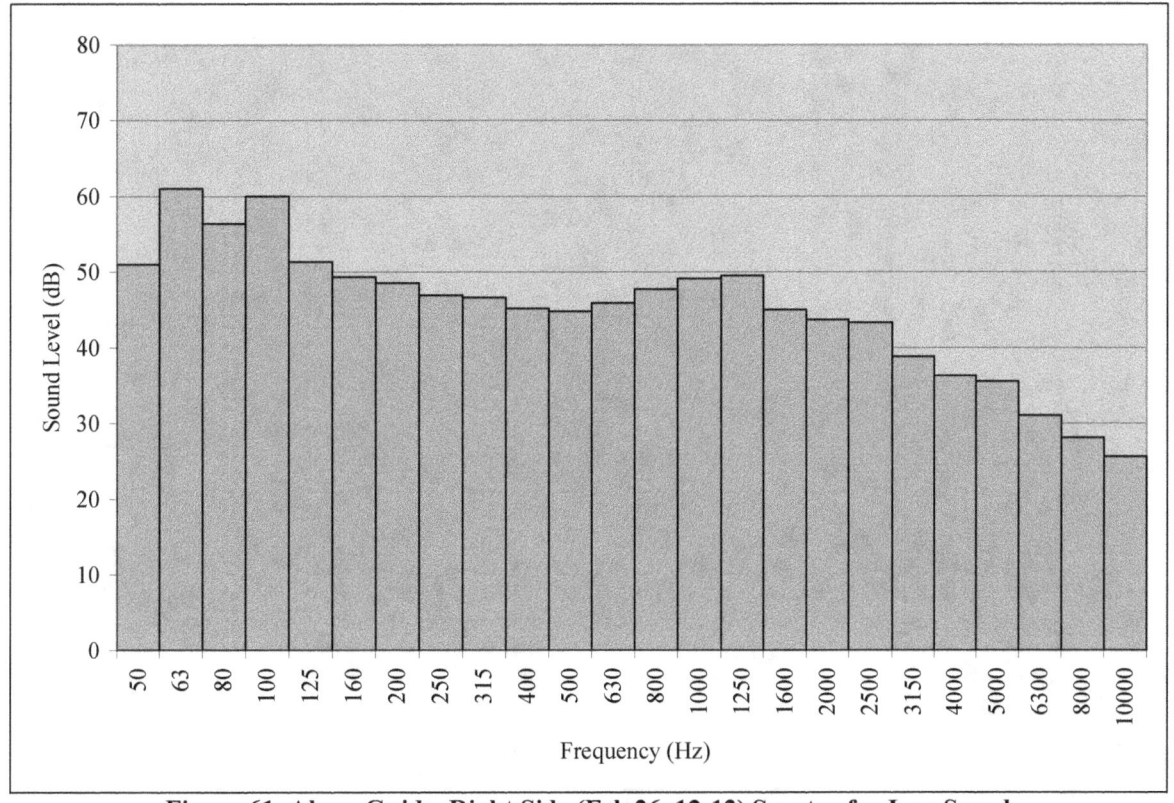

Figure 61. Alpen Guide, Right Side (Feb 26, 12:12) Spectra for Low Speed

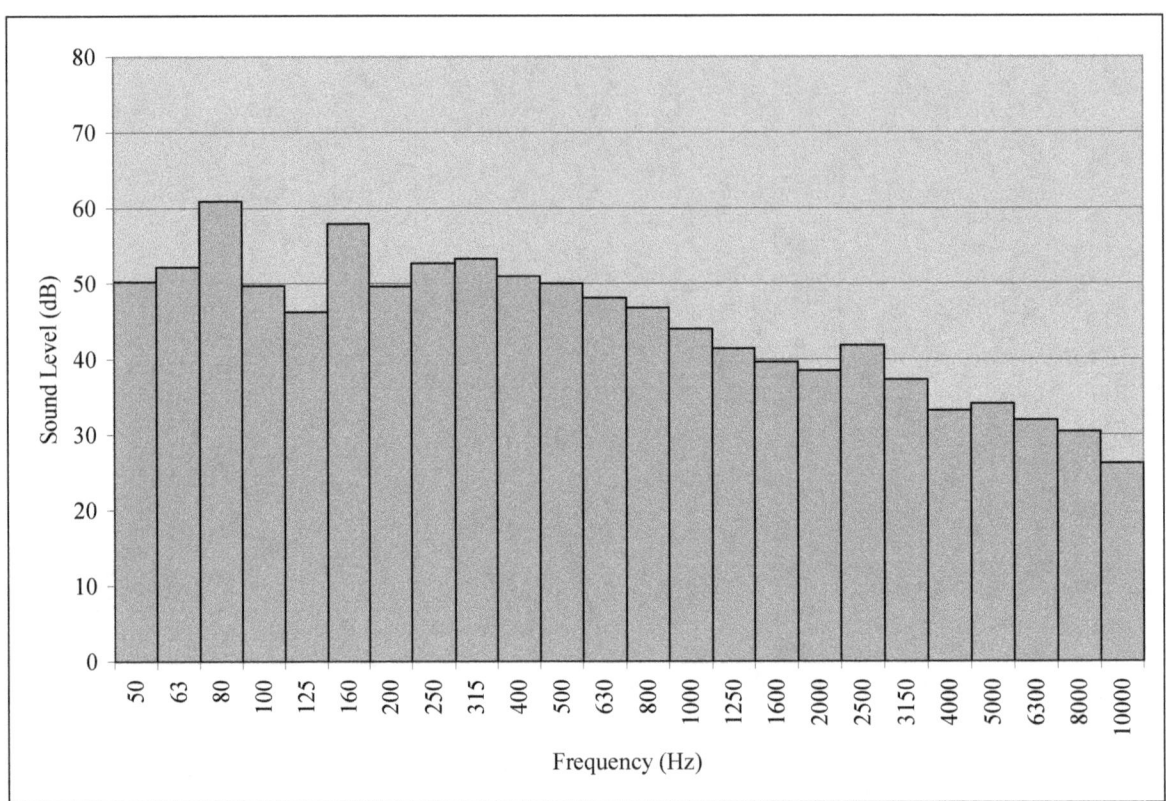

Figure 62. Rocky Mt, Right Side (Feb 26, 13:43) Spectra for Low Speed

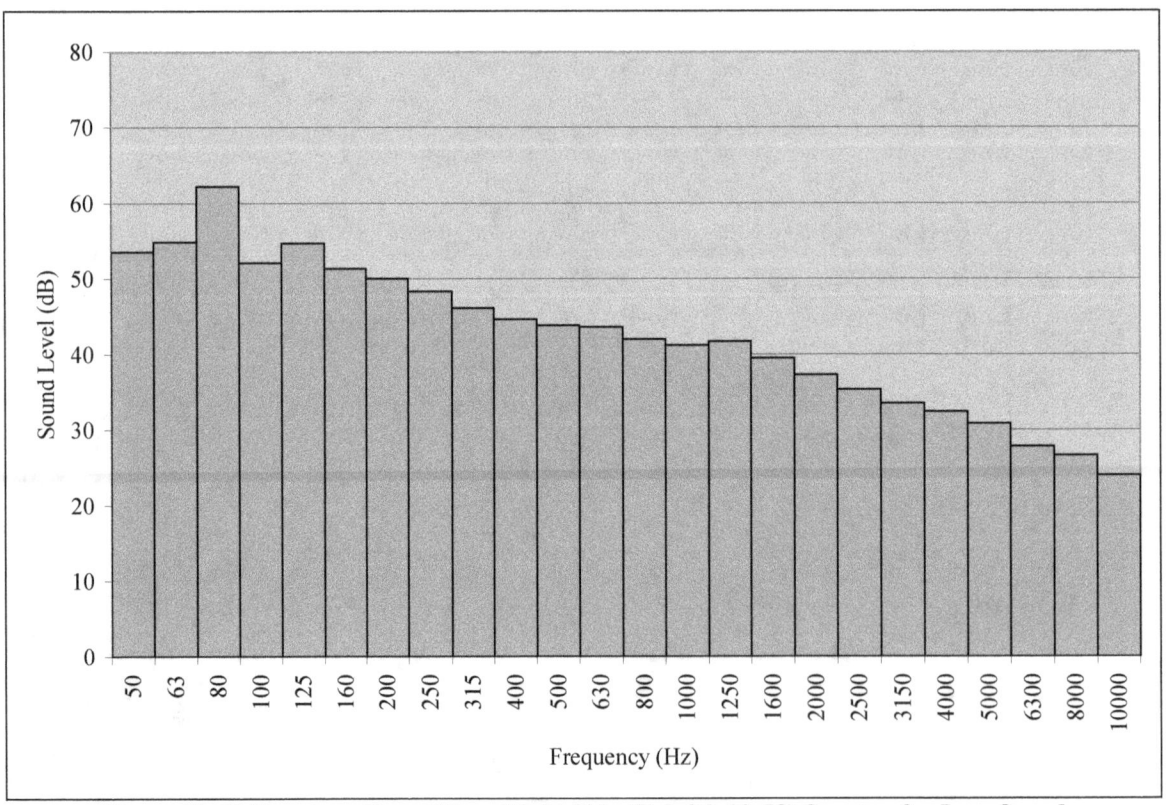

Figure 63. Yellowstone Expedition, Right Side (Feb 26, 12:53) Spectra for Low Speed

Table 27. Maximum One-Third Octave Band Levels for Select Events at High Speed (Nominally 30 mph)*, dB

Vehicle	Vehicle Side[#]	Start Time	One-Third Octave Band Center Frequency, Hz																								
			50	63	80	100	125	160	200	250	315	400	500	630	800	1000	1250	1600	2000	2500	3150	4000	5000	6300	8000	10000
Rocky Mt	Left	15:18:00	64.7	61.7	55.8	56.8	61.6	63.5	60.0	59.6	64.7	60.7	59.6	57.5	57.2	54.7	52.6	50.8	49.0	48.0	46.5	43.5	42.8	40.8	38.2	35.0
Alpen Guide	Right	14:07:00	56.8	61.2	54.5	58.0	56.2	67.5	54.7	52.0	56.3	51.5	52.3	52.6	55.4	55.3	53.9	52.4	50.9	51.0	46.9	44.1	41.2	37.4	33.7	30.0
Goose wing	Left	15:16:00	63.3	59.9	57.0	57.1	61.7	63.8	58.6	60.2	63.2	57.9	55.2	54.2	53.2	53.9	50.3	49.2	49.8	49.7	47.3	45.2	43.9	42.2	40.9	38.2
YS SC	Right	14:40:00	59.0	57.7	51.7	51.4	55.0	62.8	58.8	56.2	66.1	59.3	60.4	57.7	57.0	56.0	53.9	52.1	50.8	48.1	46.2	43.9	43.3	41.8	39.8	37.2
YS Exp	Left	14:18:00	56.8	58.8	55.8	55.3	67.9	60.7	54.7	56.0	51.0	49.6	49.3	48.0	47.3	46.9	45.6	44.4	43.1	41.1	38.7	36.4	34.7	33.4	30.9	28.6

"Left" indicates left side of vehicle is closest to the microphones. "Right" indicates right side of vehicle is closest to the microphones.

* Due to deteriorating road conditions high speed measurements were only available for the first day of measurements.

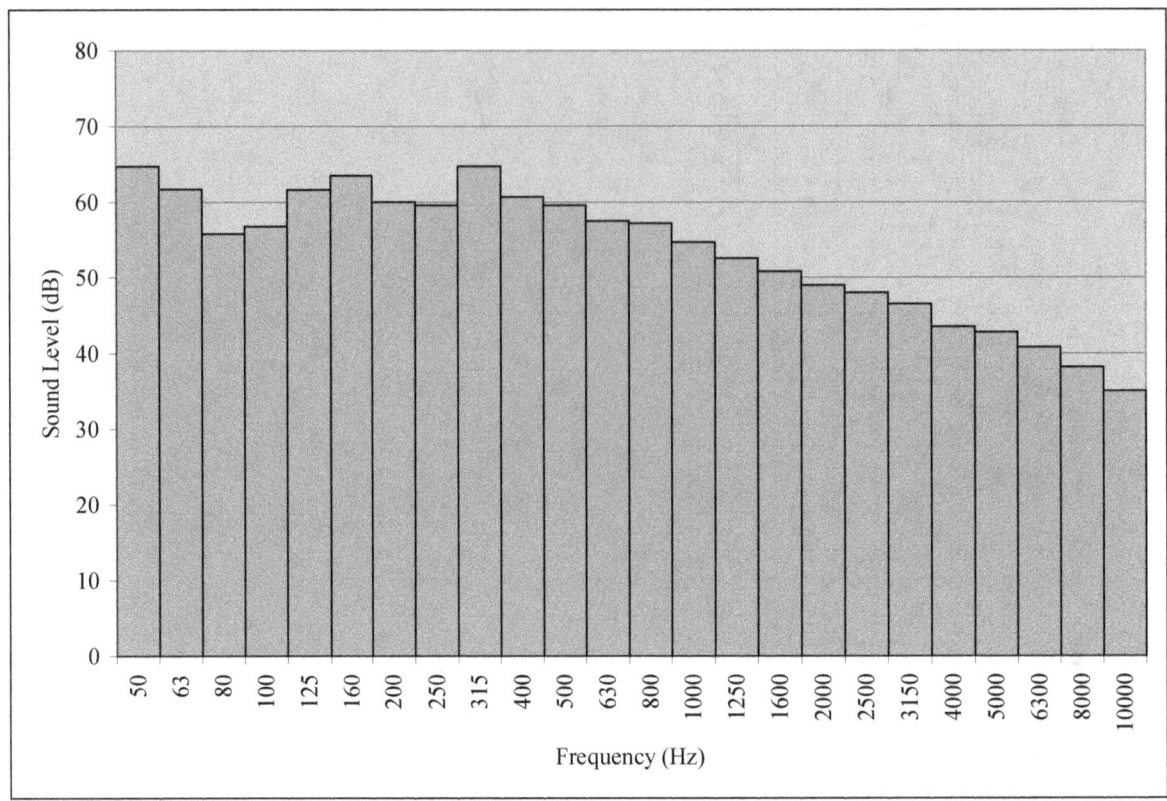

Figure 64. Rocky Mt, Left Side (Feb 26, 15:18) Spectra for High Speed

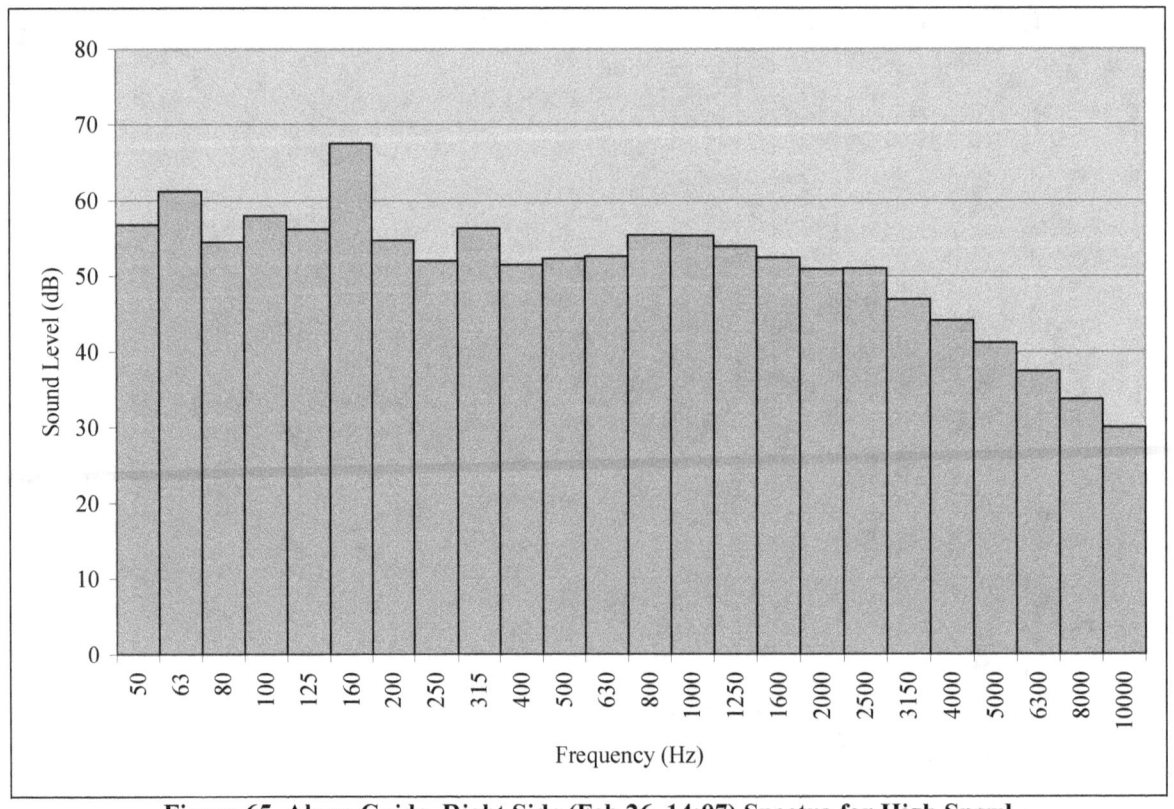

Figure 65. Alpen Guide, Right Side (Feb 26, 14:07) Spectra for High Speed

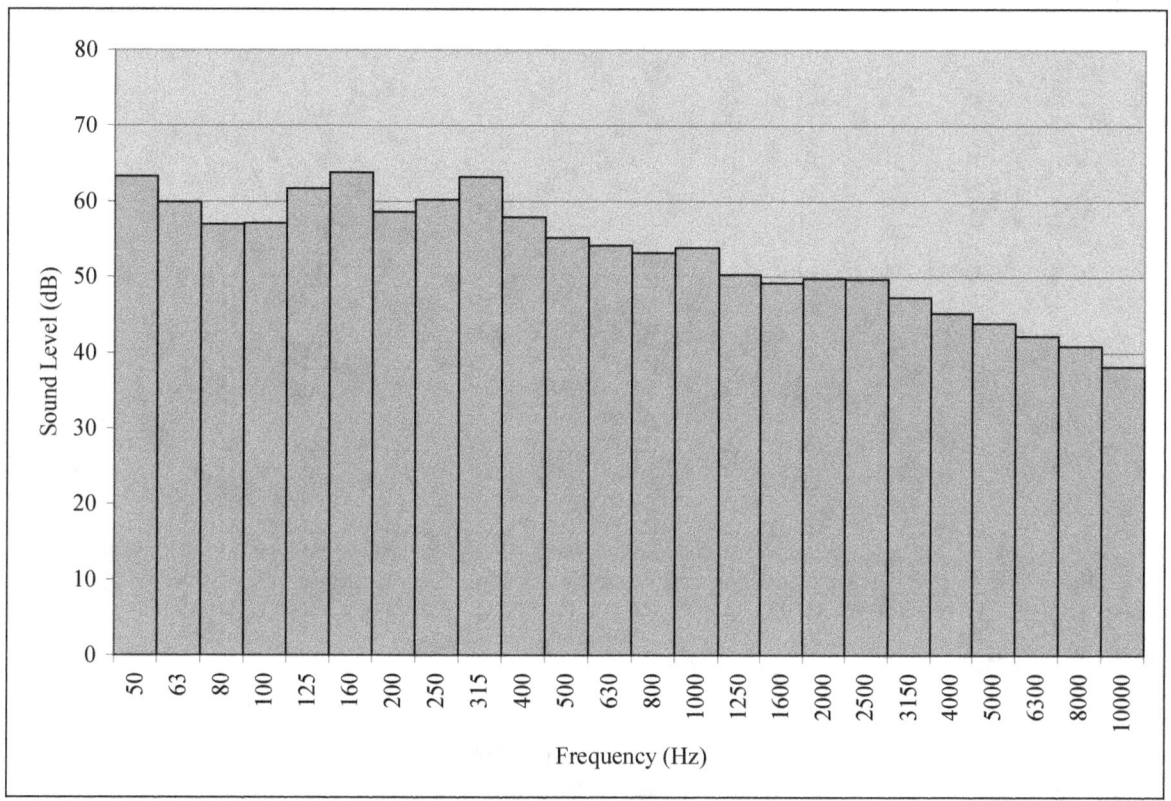

Figure 66. Goosewing Diesel Van, Left Side (Feb 26, 15:16) Spectra for High Speed

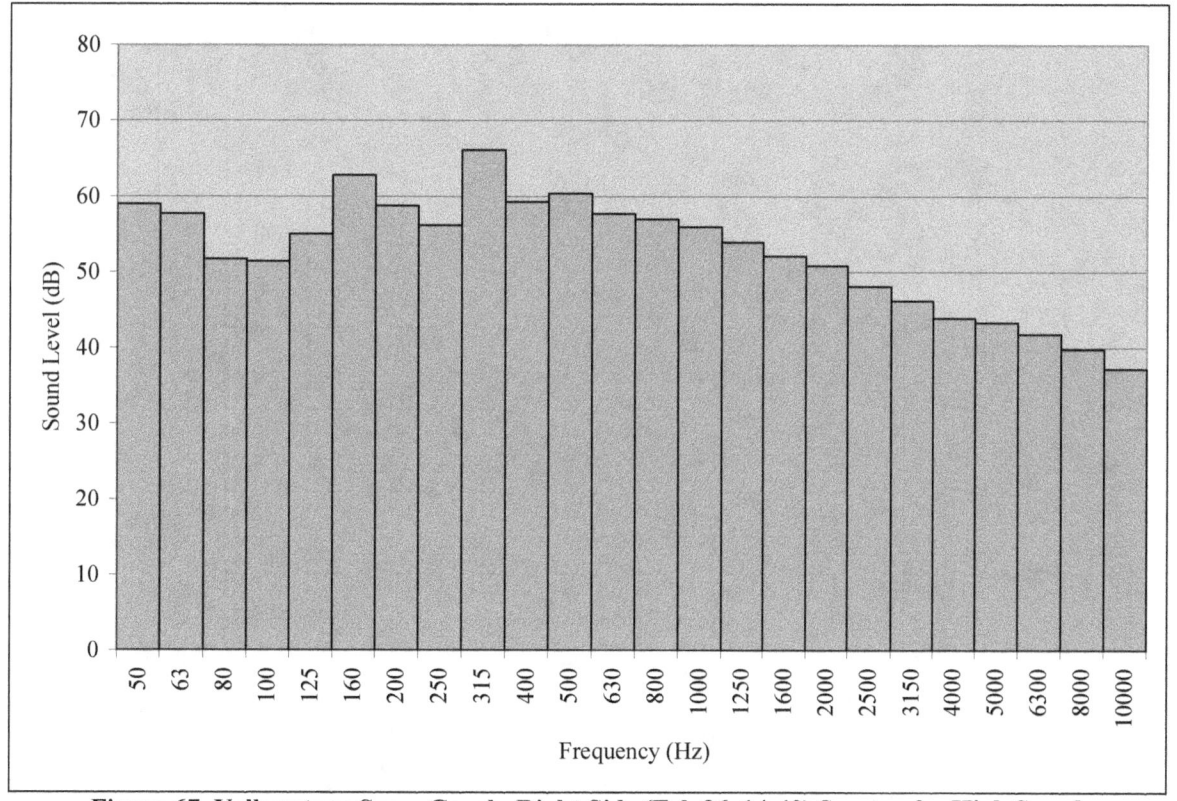

Figure 67. Yellowstone Snow Coach, Right Side (Feb 26, 14:40) Spectra for High Speed

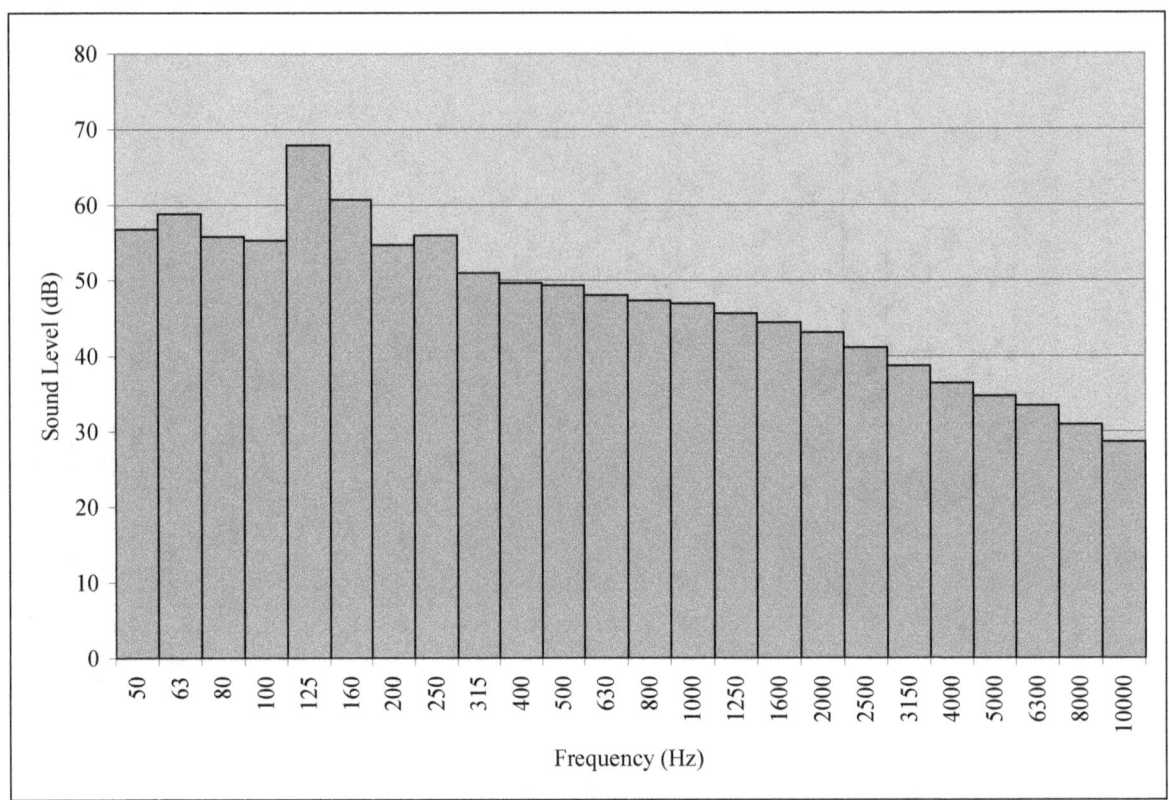

Figure 68. Yellowstone Expedition, Left Side (Feb 26, 14:18) Spectra for High Speed

References

[1] Jason Ross and Christopher Menge, "Technical Report on Noise: Winter Use Plan Final Environmental Impact Statement," HMMH Report No. 295860.18, Harris Miller Miller & Hanson Inc. 15 New England Executive Park, Burlington, MA 01803, June 2001.

[2] Christopher W. Menge and Jason C. Ross, "Draft Supplemental Report on Noise: Winter Use Plan Final Supplemental Environmental Impact Statement," HMMH Report No. 295860.400, Harris Miller Miller & Hanson Inc. 15 New England Executive Park, Burlington, MA 01803, October 2002.

[3] National Park Service, "Preliminary Draft Alternatives – Winter Use Plans," Yellowstone and Grand Teton National Parks and John D. Rockefeller, Jr., Memorial Parkway, May 19th, 2006 Draft, provided to Volpe by NPS.

[4] National Park Service, Winter Use Technical Documents, Yellowstone National Park, National Park Service, http://www.nps.gov/yell/technical/planning/winteruse/plan/, accessed on August 23, 2006.

[5] International Organization for Standardization, Committee ISO/TC 43, Acoustics, Sub-Committee SC 1, Noise, Acoustics – "Attenuation of Sound during Propagation Outdoors – Part 1: Calculation of Absorption of Sound by the Atmosphere," ISO 9513-1, Geneva, Switzerland: International Organization for Standardization, (1993).

[6] Edward J. Rickley, Gregg G. Flemming, and Christopher J. Roof, "Simplified Procedure for Computing the Absorption of Sound by the Atmosphere," Noise Control Engineering Journal, 55(6), November – December 2007.

[7] Jason Ross and Christopher Menge, "Technical Report on Noise: Winter Use Plan Final Environmental Impact Statement," HMMH Report No. 295860.18, Harris Miller Miller & Hanson Inc. 15 New England Executive Park, Burlington, MA 01803, June 2001.

[8] Christopher W. Menge and Jason C. Ross, "Draft Supplemental Report on Noise: Winter Use Plan Final Supplemental Environmental Impact Statement," HMMH Report No. 295860.400, Harris Miller Miller & Hanson Inc. 15 New England Executive Park, Burlington, MA 01803, October 2002.

[9] National Park Service, "Preliminary Draft Alternatives – Winter Use Plans," Yellowstone and Grand Teton National Parks and John D. Rockefeller, Jr., Memorial Parkway, May 19th, 2006 Draft, provided to Volpe by NPS.

[10] National Park Service, Winter Use Technical Documents, Yellowstone National Park, National Park Service, http://www.nps.gov/yell/technical/planning/winteruse/plan/, accessed on August 23, 2006.

[11] "Operational Sound Level Measurement Procedure for Snow Vehicles," Society of Automotive Engineers, SAE Surface Vehicle Recommended Practice J1161, November 1976, revised March 1983, revised April 2004.

[12] "Maximum Exterior Sound Level for Snowmobiles," Society of Automotive Engineers, SAE Surface Vehicle Recommended Practice J192, September 1970, revised March 1985, revised March 2003.

[13] Aaron L. Hastings, Gregg G. Fleming, and Cynthia S. Y. Lee, "Modeling Sound due to Over-Snow Vehicles in Yellowstone and Grand Teton National Parks," NPS-D-1201 / DOT-VNTSC-NPS-06-06, John A. Volpe National Transportation Systems Center, Acoustics Facility, Cambridge, MA 02142-1093, October 2006.

[14] Malcom J. Crocker, Ed. "Encyclopedia of Acoustics," Volume 1. John Wiley & Sons, Inc. 1997.

[15] Grant S. Anderson, Cynthia S. Y. Lee, Gregg G. Fleming, and Christopher W. Menge, "FHWA Traffic Noise Model® User's Guide," FWWA-PD-96-009/DOT-VNTSCFHWA-98-1, John A. Volpe National Transportation Systems Center, Acoustics Facility, Cambridge, MA 02142-1093, January 1998.

[16] Christopher W. Menge, Christopher F. Rossano, Grant S. Anderson, and Christopher J. Bajdek, "FHWA Traffic Noise Model® Technical Manual," FWWA-PD-96-010 / DOTVNTSC-FHWA-98-2, John A. Volpe National Transportation Systems Center, Acoustics Facility, Cambridge, MA 02142-1093, February 1998.

[17] Judith L. Rochat and Gregg G. Fleming, "Validation of FHWA's Traffic Noise Model® (TNM) Phase 1," FHWA-EP-02-031 / DOT-VNTSC-FHWA-02-01, John A. Volpe National Transportation Systems Center, Acoustics Facility, Cambridge, MA 02142-1093, August 2002.

www.ingramcontent.com/pod-product-compliance
Lightning Source LLC
Chambersburg PA
CBHW081834170526
45167CB00007B/2804